U.S. Department of Justice
Office of Justice Programs
National Institute of Justice

Citizen Review of Police

Approaches & Implementation

NIJ *Issues and Practices*

U.S. Department of Justice
Office of Justice Programs
810 Seventh Street N.W.
Washington, DC 20531

Office of Justice Programs **World Wide Web Site** *http://www.ojp.usdoj.gov*	**National Institute of Justice** **World Wide Web Site** *http://www.ojp.usdoj.gov/nij*

Citizen Review of Police: Approaches and Implementation

by Peter Finn

March 2001
NCJ 184430

National Institute of Justice

Vincent Talucci
Program Monitor

Advisory Panel*

K. Felicia Davis, J.D.
Legal Consultant and Director
 at Large
National Association for Civilian
 Oversight of Law Enforcement

Administrator
Citizen Review Board
234 Delray Avenue
Syracuse, NY 13224

Mark Gissiner
Senior Human Resources Analyst
City of Cincinnati

Immediate Past President,
 1995–99
International Association for
 Civilian Oversight of
 Law Enforcement
2665 Wayward Winds Drive
Cincinnati, OH 45230

Douglas Perez, Ph.D.
Assistant Professor
Department of Sociology
Plattsburgh State University
45 Olcott Lane
Rensselaer, NY 12144

Jerry Sanders
President and Chief
 Executive Officer
United Way of San Diego
 County
P.O. Box 23543
San Diego, CA 92193

Former Chief
San Diego Police Department

Samuel Walker, Ph.D.
Kiewit Professor
Department of Criminal Justice
University of Nebraska
 at Omaha
60th and Dodge Streets
Omaha, NE 68182

Lt. Steve Young
Vice President
Grand Lodge
Fraternal Order of Police
222 East Town Street
Columbus, OH 43215

---

*Among other criteria, advisory panel members were selected for their diverse views regarding citizen oversight of police. As a result, readers should not infer that panel members necessarily support citizen review in general or any particular type of citizen review.

Prepared for the National Institute of Justice, U.S. Department of Justice, by Abt Associates Inc., under contract #OJP–94–C–007. Points of view or opinions stated in this document are those of the author and do not necessarily represent the official position or policies of the U.S. Department of Justice.

The National Institute of Justice is a component of the Office of Justice Programs, which also includes the Bureau of Justice Assistance, the Bureau of Justice Statistics, the Office of Juvenile Justice and Delinquency Prevention, and the Office for Victims of Crime.

Foreword

In many communities in the United States, residents participate to some degree in overseeing their local law enforcement agencies. The degree varies. The most active citizen oversight boards investigate allegations of police misconduct and recommend actions to the chief or sheriff. Other citizen boards review the findings of internal police investigations and recommend that the chief or sheriff approve or reject the findings. In still others, an auditor investigates the process by which the police or sheriff's department accept or investigate complaints and reports to the department and the public on the thoroughness and fairness of the process.

Citizen oversight systems, originally designed to temper police discretion in the 1950s, have steadily grown in number through the 1990s. But determining the proper role has a troubled history.

This publication is intended to help citizens, law enforcement officers and executives, union leaders, and public interest groups understand the advantages and disadvantages of various oversight systems and components.

In describing the operation of nine very different approaches to citizen oversight, the authors do not extol or disparage citizen oversight but rather try to help jurisdictions interested in creating a new or enhancing an existing oversight system by:

• Describing the types of citizen oversight.

• Presenting programmatic information from various jurisdictions with existing citizen oversight systems.

• Examining the social and monetary benefits and costs of different systems.

The report also addresses staffing; examines ways to resolve potential conflicts between oversight bodies and police; and explores monitoring, evaluation, and funding concerns.

No one system works best for everyone. Communities must take responsibility for fashioning a system that fits their local situation and unique needs. Ultimately, the author notes, the talent, fairness, dedication, and flexibility of the key participants are more important to the procedure's success than is the system's structure.

Acknowledgments

I thank the many individuals who patiently answered my questions and sent me materials about their citizen oversight procedures. In particular, I thank the following oversight directors and coordinators: Barbara Attard, Lisa Botsko, Mary Dunlap, Suzanne Elefante, Patricia Hughes, Liana Perez, Melvin Sears, Todd Samolis, Ruth Siedschlag, and Joseph Valu.

The following advisory panel members (whose titles are listed on the back of the title page) provided a large number of helpful comments during a 1-day meeting in Washington, D.C., and reviewed the draft report: K. Felicia Davis, Mark Gissiner, Douglas Perez, Jerry Sanders, Samuel Walker, and Steve Young. Among other criteria, advisory panel members were selected for their diverse views regarding citizen oversight of police. As a result, readers should not infer that the panel members necessarily support citizen review in general or any particular type of citizen review.

Benjamin Tucker, former Deputy Director of Operations of the Office of Community Oriented Policing Services, and Phyllis McDonald, Social Science Analyst with the National Institute of Justice (NIJ), also participated in the board meeting and made important contributions. Pierce Murphy, Community Ombudsman in Boise, Idaho, provided valuable suggestions for improving the report.

Vincent Talucci, Program Manager for the project at NIJ, provided wise guidance and constant support. Terence Dunworth, Managing Vice President at Abt Associates Inc., offered numerous suggestions for improving the report, most important, a complete reconfiguration of the executive summary and discussion of program costs. Mary-Ellen Perry and Joan Gilbert carefully produced the numerous report drafts.

Peter Finn
Associate
Abt Associates Inc.

Executive Summary

Introduction

There has been a considerable increase in the number of procedures involving citizen oversight of police implemented by cities and counties in the 1990s. However, many of these procedures have had a troubled history involving serious—even bitter—conflict among the involved parties. *Citizen Review of Police: Approaches and Implementation* is designed to help jurisdictions that may decide to establish—or wish to improve—an oversight system to avoid or eliminate these battles. At the same time, the publication can help oversight planners understand and choose among the many options available for structuring a citizen review procedure. Finally, current oversight staff and volunteers may find it useful to review the publication as a way of learning more about the field.

To provide this assistance, *Citizen Review of Police* describes the operations of nine very different systems of citizen oversight. However, the publication does not promote any particular type of citizen review—or citizen review in general. Rather, the report is intended to help local government executives and legislators, as well as police and sheriff's department administrators, union leaders, and local citizen groups and public interest organizations, learn about the advantages, drawbacks, and limitations of a variety of oversight systems and components.

Types of Citizen Oversight

There is no single model of citizen oversight. However, most procedures have features that fall into one of four types of oversight systems:

- Type 1: *Citizens investigate allegations* of police misconduct and *recommend findings* to the chief or sheriff.

- Type 2: Police officers investigate allegations and develop findings; *citizens review and recommend* that the chief or sheriff approve or reject the findings.

- Type 3: Complainants may *appeal findings* established by the police or sheriff's department *to citizens,* who review them and then recommend their own findings to the chief or sheriff.

- Type 4: An auditor *investigates the process* by which the police or sheriff's department accepts and investigates complaints and reports on the thoroughness and fairness of the process to the department and the public.

All four types of oversight are represented among the nine citizen review systems described in this report (see exhibit 1).

Each type of system has advantages and drawbacks. For example, oversight systems that involve investigating citizen complaints (type 1) can help reassure the public that investigations of citizen complaints are thorough and fair. However, hiring professional investigators can be expensive, and the investigations model typically has no mechanism for soliciting the public's general concerns about police conduct.

Whatever their specific advantages, any type of citizen oversight needs to be part of a larger structure of internal and external police accountability; citizen oversight alone cannot ensure that police will act responsibly.

Oversight Costs

Exhibit 2 presents the nine oversight systems arranged in ascending order of budget levels along with their activity levels for 1997. As shown, there is a theoretical relationship between the four *types* of oversight systems and cost.

- Type 1 oversight systems, in which citizens investigate allegations and recommend findings (Berkeley, Flint, Minneapolis, San Francisco), are the most expensive largely because professional investigators must be hired to conduct the investigations—lay citizens do not have the expertise or the time.

Exhibit 1. Type and Selected Features of Nine Oversight Systems

System	Type*	Openness to Public Scrutiny	Mediation Option	Subpoena Power	Officer Legal Representation
Berkeley Police Review Commission (PRC)	1	• hearings and commission decisions open to public and media • general PRC meetings available for public to express concerns • full public report, including interview transcripts • city manager makes response public after review of PRC and internal affairs (IA) findings • appeal process • IA's dispositions and discipline not public	dormant	yes	during investigation; during hearing
Flint Office of the Ombudsman	1	• findings distributed to media and city archives • no appeal • chief's finding public, but not discipline	no	yes, but never used	not interviewed in person
Minneapolis Civilian Police Review Authority (CRA)	1	• hearings are private • general public invited to monthly CRA meeting to express concerns • appeal process • complainant told whether complaint was sustained • chief's discipline not public until final disposition	yes	no, but cooperation required under *Garrity* ruling	during investigation, union representative may advise officer; during hearing, union attorney defends officer
Orange County Citizen Review Board	2	• hearings open to public and media scrutiny • findings and the sheriff's discipline are matters of public record • no appeal	no	yes, but never used	during hearings
Portland Police Internal Investigations Auditing Committee (PIIAC)	3, 4	• PIIAC audits open to public and media • citizen advisory subcommittee meetings open to public and media • appeal to city council • PIIAC decisions are public; chief's discipline is not	no	yes	none
Rochester Civilian Review Board	2	• reviews are closed • results are not public • no appeal	yes	no	none
St. Paul Police Civilian Internal Affairs Review Commission	2	• hearings are closed • no appeal • no publicizing of disciplinary recommendations	no	yes, but never used	none
San Francisco Office of Citizen Complaints	1	• chief's hearings are closed • police commission hearings are public • appeal process for officers • complaint histories and findings confidential • chief's discipline not public	yes	yes	during investigation; during hearing
Tucson Independent Police Auditor and Citizen Police Advisory Review Board	2, 4	• monitoring is private • appeal process • board holds monthly public meeting at which public may raise concerns	no	no	not applicable

* Type 1: citizens investigate allegations and recommend findings; type 2: police officers investigate allegations and develop findings; citizens review findings; type 3: complainants appeal police findings to citizens; type 4: an auditor investigates the police or sheriff's department's investigation process.

Exhibit 2. 1997 Oversight System Costs in Relation to Responsibilities and Activity

Type	Location	Name	Budget	Paid Staff	Activity	Mean Cost per Complaint Filed and Investigated
2	Orange County pop: 749,631 sworn: 1,134	Citizen Review Board	$20,000 (1/5 time)	2 part time	45 board hearings } 45	$444
2	St. Paul pop: 259,606 sworn: 581	Police Civilian Internal Affairs Review Commission	$37,160[a]	1 part time	71 cases reviewed } 71	$523
3, 4	Portland pop: 480,824 sworn: 1,004	Police Internal Investigations Auditing Committee	$43,000	1 full time	21 appeals processed 98 audits of completed cases } 119	$361
2	Rochester pop: 221,594 sworn: 685	Civilian Review Board	$128,069	1 full time 3 part time	26 cases reviewed 4 cases mediated } 30	$4,269
2, 4	Tucson pop: 449,002 sworn: 865	Independent Police Auditor and Citizen Police Advisory Review Board	$144,150[b]	2 full time	289 citizen contacts (9/1/97–6/30/98) 96 complaints filed with auditor 63 investigations monitored } 159	$755[c]
1	Flint pop: 134,881 sworn: 333	Office of the Ombudsman	$173,811[d]	2 full time[e] 1 part time	313 cases against police investigated (1996) } 313	$555
1	Berkeley pop: 107,800 sworn: 190	Police Review Commission	$277,255	4 full time	42 complaints filed and investigated 34 cases closed for lack of merit or complainant cooperation 12 hearings } 57[f]	$4,864[g]
1	Minneapolis pop: 358,785 sworn: 919	Civilian Police Review Authority	$504,213	7 full time	715 citizen contacts 159 formal complaints investigated } 159 14 cases mediated 6 stipulations 3 hearings	$3,171[h]
1	San Francisco pop: 735,315 sworn: 2,100	Office of Citizen Complaints	$2,198,778	30 full time	1,126 complaints received 983 cases investigated and closed } 983 50 chief's hearings 6 police commission trials	$2,237

a. Because the board has never spent its $10,000 allocation for hiring an independent investigator, its true budget has been $27,160.
b. Includes a one-time startup cost of $20,000 for remodeling office.
c. Based on a 10-month prorated budget of $120,058.
d. This represents the proportion of the staff who handle complaints against the police.
e. Two of seven investigators handle complaints full time against the police; a third investigator spends one-quarter time handling police complaints.
f. Because some of the 42 cases filed and investigated were among the 34 cases closed in 1997, the total activity does not represent the sum of the 42 complaints investigated and the 34 cases closed.
g. The mean cost per complaint filed for 1997 was atypically high because, for example, oversight staff conducted a major pepper spray study, staffed a medical marijuana task force, and held a major public hearing.
h. This figure is misleadingly high as a measure of overall oversight costs because it does not take into consideration the 556 other contacts authority staff had with the public that did not result in formal complaints.

Executive Summary

- Type 2 systems, in which citizens review the internal affairs unit's findings (e.g., Orange County, Rochester, St. Paul), tend to be inexpensive because volunteers typically conduct the reviews.

- Type 3 systems, in which citizens review complainants' appeals of police findings (Portland), can also be inexpensive because of the use of volunteers.

- Type 4 systems, in which auditors inspect the police or sheriff's department's own complaint investigation process (Portland, Tucson), tend to fall in the midlevel price range. On one hand, like type 1 systems, only a paid professional has the expertise and time to conduct a proper audit. On the other hand, typically only one person needs to be hired because the auditing process is less time consuming than conducting investigations of citizen complaints.

In practice, however, there is an inconsistent relationship between oversight type and cost. This is because, when examined closely, many oversight operations are not "pure" examples of a type 1, 2, 3, or 4 system. For example, two jurisdictions have combined two different oversight approaches: Portland has a citizen appeals board (type 3) and an auditor who monitors the police bureau's complaint investigation process (type 4); Tucson has both a citizen board that reviews internal affairs findings (type 2) and an auditor (type 4). Consequently, the actual cost for a given type of oversight system may be more or less expensive than the cost of a pure type. Furthermore, each type of oversight system can incorporate features that may increase or decrease its expenses, ranging from providing policy recommendations to a mediation option. The choice of staffing option also will affect expenditures, including using volunteer staff or in-kind services and materials, hiring paid staff, or diverting part of the time of an existing city or police employee to oversight functions. As a result, it is difficult to predict an oversight system's actual costs before determining all its features and activities.

Finally, more money may not buy more oversight activity or increase use of the system—that is, boost the number of complaints, hearings, mediations, policy recommendations, reviews, or audits. A variety of cost-insensitive considerations—the public's perception of the system's fairness, the director's impartiality and talent, the level of cooperation from the police or sheriff's department, and restrictions on the kinds of complaints the system will be prohibited from handling or required to accept—can prevent additional funds from resulting in increased use of the oversight system. That said, an oversight procedure that is underfunded will not only have difficulty achieving its objectives, it also may create more controversy surrounding police accountability than it resolves.

Conclusions

This report suggests at least four other significant conclusions regarding citizen oversight of the police.

Local jurisdictions that wish to establish citizen review have to take on the responsibility to make difficult choices about the type of oversight system they should fashion. The tremendous variation in how the nine oversight systems described in this report conduct business—and pay for their activities—may seem discouraging: The lack of similarity makes it difficult for other jurisdictions to make an automatic selection of commonly implemented citizen review features around which they can structure their own oversight procedures. On the positive side, this diversity means jurisdictions do not have to feel obligated to follow slavishly any one model or approach; they have the freedom to tailor the various components of their system to the particular needs and characteristics of their populations, law enforcement agencies, statutes, collective bargaining agreements, and pressure groups.

Many individuals and groups believe that citizen oversight, despite its serious limitations, can have important benefits. Complainants have reported that they:

- Feel "validated" when the oversight body agrees with their allegations—or when they have an opportunity to be heard by an independent overseer regardless of the outcome.

- Are satisfied at being able to express their concerns in person to the officer.

- Feel they are contributing to holding the department accountable for officers' behavior.

Police and sheriff's department administrators have reported that citizen oversight:

- Improves their relationship and image with the community.

- Has strengthened the quality of the department's internal investigations of alleged officer misconduct and reassured the public that the process is thorough and fair.

- Has made valuable policy and procedure recommendations.

Local elected and appointed officials say an oversight procedure:

- Enables them to demonstrate their concern to eliminate police misconduct.

- Reduces in some cases the number of civil lawsuits (or successful suits) against their cities or counties.

It is sometimes possible to overcome disagreements between oversight operations and police and sheriff's departments. The report identifies many points of conflict between oversight systems and police and sheriff's departments—and with officer unions. However, as illustrated in exhibit 3, there are positions each side can take and explanations it can offer that can sometimes make the system acceptable to everyone involved. A critical step to minimizing conflict is for the police or sheriff's department—and union leadership—to act as colleagues in the planning process.

The talent, fairness, dedication, and flexibility of the key participants—in particular, the oversight system's director, chief elected official, police chief or sheriff, and union president—are more important to the procedure's success than is the system's structure. The report identifies jurisdictions in which these individuals have worked together cooperatively. An effective procedure for selecting competent and objective oversight investigators, board members, and administrators—and for training them thoroughly—is also critical for the oversight procedure to thrive.

Exhibit 4 is a checklist oversight system planners can consult to help identify some of the decisions they will have to make in designing and setting up a new or revised review procedure. The exhibit indicates where in this report's text each decision is discussed.

Exhibit 3. Concerns Many Police and Sheriff's Departments—and Union Leaders—Express About Citizen Oversight—and Possible Responses

Assertion: Citizens Should Not Interfere in Police Work

Concerns	Responses
• The chief must be held accountable for discipline to prevent misconduct.	• Most oversight bodies are only advisory.
• Internal affairs already does a good job.	• Even when the department already imposes appropriate discipline without citizen review, an oversight procedure can reassure skeptical citizens that the agency is doing its job in this respect. • The next chief or sheriff may not be as conscientious about ensuring that the department investigates complaints fairly and thoroughly.

Assertion: Citizens Do Not Understand Police Work

Concerns	Responses
• Oversight staff lack experience in police work.	• Board members typically have pertinent materials available for review, and ranking officers are usually present during hearings to explain department procedures. • Oversight administrators need to describe the often extensive training they and their staff receive. • Citizen review is just that—*citizens* reviewing police behavior as private citizens.
• Only physicians review doctors, and only attorneys review lawyers.	• Doctors and lawyers have been criticized for doing a poor job of monitoring *their* colleagues' behavior.

Assertion: The Process Is Unfair

Concerns	Responses
• Oversight staff may have an "agenda"—they are biased against the police.	• Oversight staff need to inform the department when they decide in officers' favor. • Oversight staff and police need to meet to iron out misconceptions and conflict.
• Not sustained findings remain in officers' files.	• Indecisive findings are unfair to both parties and should therefore be reduced in favor of unfounded, exonerated, or sustained findings.
• Adding allegations unrelated to the citizen's complaint is unfair.	• Internal affairs units themselves add allegations in some departments.
• Some citizens use the system to prepare for civil suits.	• Board findings can sometimes help officers and departments defend against civil suits.

Exhibit 4. Decisions Oversight Planners Need to Make

Decision	Discussion in Text
Establish a Planning or Advisory Group	
Identify the key actors	N/A
Establish a formal planning committee	N/A
Identify sources of technical assistance	chapter 8
Plan for Monitoring and Evaluation	chapter 7
Design a monitoring plan	
Design an evaluation plan	
identify program's objectives	
select an evaluator	
develop measures of effectiveness	
develop measurement methods	
collect data	
analyze data	
interpret and report findings	
Select Program Type	
Review existing program models and materials	chapter 3
Visit selected programs to interview staff and observe procedures	N/A
Identify tradeoffs involved in different components	chapter 1
Consider Taking on Other Oversight Responsibilities	chapter 3
Provide policy recommendations	
Offer mediation	
Assist with early warning system	
Determine Outreach Methods	chapter 5
Establish Extent of Openness	chapter 5
Public or private hearings	
Reporting procedures	
type	
content	
frequency	
distribution	
Identify Staffing Needs	chapter 4
Decide on type and number of staff	
volunteer board members	
paid investigators	
director/ombudsman/auditor	
use existing staff	
hire new staff	
other staff (support, management information system)	
Determine how to recruit, screen, and train staff	
Select Program Structure	chapter 5
Establish eligibility criteria for complainants	
Identify types of cases to review or investigate	
Decide where complainants may file	
at police station or sheriff's department	
at oversight program	
other (city hall, etc.)	
Consider whether to seek subpoena power	
Develop timelines for completing each phase of the complaint process	
Develop plan for minimizing delays in case processing	
Estimate Budget Needs	chapter 7

N/A = not applicable

Contents

Foreword iii

Acknowledgments v

Executive Summary vii
    Introduction vii
    Types of Citizen Oversight vii
    Oversight Costs vii
    Conclusions x

Chapter 1: Introduction 1
    Key Points 1
    What the Publication Is Intended to Do 2
        Audiences and purposes for *Citizen Review of Police* 2
        The need for the report 3
    Features of the Report 4
        Sources of information for the publication 4
        Terminology used in the report 5
    Types of Citizen Review 6
    Potential Benefits of Citizen Oversight 6
        Potential benefits to complainants 7
        Potential benefits to police and sheriff's departments 8
        Potential benefits to elected and appointed officials 10
        Potential benefits to the community at large 12
    Limitations to Citizen Review 12
    Notes 15

Chapter 2: Case Studies of Nine Oversight Procedures 17
    The Berkeley, California, Police Review Commission: A Citizen Board and the Police
        Department Investigate Complaints Simultaneously 21
        Background 21
        The review process 21
        Other activities 24
        Staffing and budget 25
        Distinctive features 25

Contents

The Flint, Michigan, Ombudsman's Office: An Ombudsman Investigates Selected Citizen
Complaints Against All City Departments and Agencies 26
    Background 26
    The review process 27
    Other activities 28
    Staffing and budget 28
    Distinctive features 30

The Minneapolis Civilian Police Review Authority: An Oversight System Investigates
and Hears Citizen Complaints 30
    Background 30
    The complaint process 32
    Other CRA activities 34
    Staffing and budget 35
    Distinctive features 36

The Orange County, Florida, Citizen Review Board: A Sheriff's Department
Provides Executive Support to an Independent Review Board 37
    Background 37
    The CRB procedure 37
    Other activities 39
    Staffing and budget 40
    Distinctive features 40

The Portland, Oregon, Police Internal Investigations Auditing Committee: A City Council,
Citizen Advisers, and a Professional Examiner Share Oversight Responsibilities 41
    Background 41
    Citizen appeals of IA findings 42
    Audits 43
    PIIAC (city council and mayor) 44
    Other activities 45
    Staffing and budget 45
    Distinctive features 45

The Rochester, New York, Civilian Review Board: Trained Mediators Review Citizen Complaints 46
    Background 46
    Procedure 47
    Other responsibilities 49
    Staffing and budget 49
    Distinctive features 50

The St. Paul Police Civilian Internal Affairs Review Commission: A Police-Managed
Board Recommends Discipline 51
    Background 51
    The review process 52
    Staffing and budget 54
    Distinctive features 55

San Francisco's Office of Citizen Complaints: An Independent Body Investigates
   Most Citizen Complaints for the Police Department 55
    Background 55
    The complaint filing process 56
    Other activities 60
    Staffing and budget 60
    Distinctive features 60

Tucson's Dual Oversight System: A Professional Auditor and a Citizen
   Review Board Collaborate 62
    Background 62
    The independent police auditor 62
    Citizen Police Advisory Review Board 65
    The relationship between the board and the auditor 66
    Distinctive features 66

Chapter 3: Other Oversight Responsibilities 69
Key Points 69
Policy Recommendations 69
    The process of developing policy recommendations 70
    Examples of policy recommendations 71
Mediation 72
    Formal mediation: The process 72
    Potential benefits of mediation 74
    Drawbacks to mediation 80
Early Warning Systems 80
    Oversight involvement in EWS 81
    Benefits and drawbacks of EWS 81
Notes 82

Chapter 4: Staffing 83
Key Points 83
Volunteer Board Members 84
    Planning decisions 84
    Selection of board members 85
    Recruitment 86
    Training 87
    Inservice training 88
Investigators 88
    Recruitment 89
    Training 90
Executive Director or Auditor 90
Notes 90

Contents

Chapter 5: Addressing Important Issues in Citizen Oversight 93
    Key Points 93
    Outreach 94
        Publicity materials 94
        Postings 96
        Media 96
        Neighborhood groups and other agencies 96
        Filing locations 96
        Referrals by the police 97
    Issues of Oversight Mechanics 98
        Oversight's legal basis 98
        Eligible complainants and cases 98
        Subpoena power 99
        Other structural issues 101
    Minimizing Delays 101
    Openness of Oversight Proceedings 103
    Politics 104
        Conflict among local government officials 104
        Agendas on the part of volunteers 105
    Notes 105

Chapter 6: Resolving Potential Conflicts Between Oversight Bodies and Police 107
    Key Points 107
    Preliminary Steps for Minimizing Conflict 108
    Police Criticisms of Oversight Procedures 109
        Citizens should not interfere with police work 109
        Citizens do not understand police work 111
        The process is unfair 112
    Oversight Criticisms of the Police 116
    Working With the Union 117
        Historical conflict between most police unions and citizen oversight bodies 117
        Approaches to collaboration 118
        Time may help but is not the cure-all 120
    Notes 121

Chapter 7: Monitoring, Evaluation, and Funding 123
    Key Points 123
    Monitoring 124
        The intake process 124
        Investigators' work 124
        Board members' performance 125

Consumer Satisfaction Surveys 125
Evaluating the Citizen Oversight Process 127
    Establishing objectives 127
    The Albuquerque evaluation 128
Funding 128
    The relationship between oversight costs and oversight activity 128
    The relationship between cost and oversight type 134
    The importance of adequate funding 135
Notes 136

Chapter 8: Additional Sources of Help 137
Key Points 137
Organizations 137
Selected Program Materials 137
Selected Publications and Reports 139
Individuals With Experience in Citizen Oversight of Police 140

Glossary 143

Appendixes
Appendix A: Rochester Internal Affairs Request for Board Member Evaluations
  of Investigators' Investigations 145
Appendix B: Portland Auditor's Guidelines for Reviewing Internal Affairs Investigations 147
Appendix C: San Francisco Complaint Intake Form 151
Appendix D: San Francisco Policy for Citizen Monitoring of Police During Demonstrations 155
Appendix E: Cost Estimates for Different Modifications to Tucson's Existing
  Oversight Procedure 163

Index 165

List of Exhibits
Exhibit 1. Type and Selected Features of Nine Oversight Systems viii
Exhibit 2. 1997 Oversight System Costs in Relation to Responsibilities and Activity ix
Exhibit 3. Concerns Many Police and Sheriff's Departments—and Union Leaders—Express
  About Citizen Oversight—and Possible Responses xii
Exhibit 4. Decisions Oversight Planners Need to Make xiii
Exhibit 1–1. Sample Tradeoffs Jurisdictions Need to Consider in Choosing an Oversight Procedure ...7
Exhibit 1–2. Potential Benefits of Citizen Oversight for Police and Sheriff's Departments 8
Exhibit 1–3. Limitations to Citizen Oversight 13

CONTENTS

Exhibit 2–1. Selected Features of the Nine Oversight Systems 18
Exhibit 2–2. Additional Features of the Nine Oversight Systems 19
Exhibit 2–3. Citizen Complaint Process in Berkeley 22
Exhibit 2–4. Berkeley Police Review Commission Budget, Fiscal Year 1998–99 25
Exhibit 2–5. Flint Office of the Ombudsman's Investigation Process 27
Exhibit 2–6. Flint Ombudsman's Office 1998–99 Budget 30
Exhibit 2–7. Disposition of 159 Signed Complaints in 1997 32
Exhibit 2–8. Minneapolis Civilian Police Review Authority Complaint Process 32
Exhibit 2–9. Minneapolis Civilian Police Review Authority 1998 Budget 36
Exhibit 2–10. The Orange County Citizen Review Process 38
Exhibit 2–11. Steps in the Portland Audit Procedure 42
Exhibit 2–12. Rochester Citizen Oversight Process 47
Exhibit 2–13. Center for Dispute Settlement CRB and Conciliation Budget for
             Fiscal Year 1998–99 50
Exhibit 2–14. St. Paul Citizen Review Process 53
Exhibit 2–15. Police Civilian Internal Affairs Review Commission 1995 Budget 55
Exhibit 2–16. San Francisco's Oversight Process 57
Exhibit 2–17. Office of Citizen Complaints 1998–99 Budget 61
Exhibit 2–18. Citizen Oversight Process in Tucson 63
Exhibit 2–19. Tucson Independent Police Auditor Budgets for Fiscal Years 1997–98
             and 1998–99 65
Exhibit 3–1. Portland Police Bureau Bulletin on Handcuffing Issued in Response
             to Auditor's Recommendation 73
Exhibit 3–2. Minneapolis Mediation Program Rules 75
Exhibit 3–3. Mediation Summary and Agreement Form 76
Exhibit 3–4. Potential Benefits of Mediation to Citizens and Police 77
Exhibit 5–1. Oversight Outreach Methods 96
Exhibit 5–2. OCC Incident Information Card 98
Exhibit 5–3. Flier Included in Letter the Portland Police Bureau Sends to Complainants
             Notifying Them of Their Cases/Findings 99
Exhibit 5–4. Number of Days Each of 10 Complaints Remained at 3 San Jose Police
             Department Offices 102
Exhibit 6–1. Concerns Many Police and Sheriff's Departments—and Union Leaders—
             Express About Citizen Oversight—and Possible Responses 110
Exhibit 7–1. Minneapolis Consumer Satisfaction Post-Outcome Survey 126
Exhibit 7–2. Costs of Nine Oversight Systems in 1997 in Relation to Responsibilities
             and Activity 131
Exhibit 8–1. Individuals With Experience in Citizen Oversight of Police 141

Chapter 1: Introduction

KEY POINTS

- *Citizen Review of Police: Approaches and Implementation* is written primarily for local government officials and legislators. Union leaders, local citizen groups, and new oversight staff may also find the publication useful.

- The publication describes nine citizen oversight procedures to enable these audiences to benefit from the experiences of communities that have already established oversight procedures.

- While there is no single model of citizen oversight, most systems fall into one of four types:

  — Type 1: *Citizens investigate allegations* of police misconduct and *recommend findings* to the chief or sheriff.

  — Type 2: Police officers investigate allegations and develop findings; *citizens review and recommend* that the chief or sheriff approve or reject the findings.

  — Type 3: Complainants may *appeal findings* established by the police department *to citizens,* who review them and then recommend their own findings to the chief or sheriff.

  — Type 4: An auditor *investigates the process* by which the police or sheriff's department accepts and investigates complaints and reports on the process' thoroughness and fairness.

- Oversight bodies can also:

  — Recommend changes in department policies and procedures and suggest improvements in training.

  — Arrange for mediation.

  — Assist the police or sheriff's department to develop or operate an early warning system for identifying problem officers.

- If they wish to implement citizen review, to make an informed decision about which type of oversight procedure to adopt jurisdictions need to examine tradeoffs inherent in choosing a model: Most features of every model have drawbacks as well as benefits.

- Citizen oversight has the potential to benefit many groups.

- Complainants have reported feeling:

  — "Validated" when their allegations are sustained—or merely appreciated having an opportunity to be heard by an independent third party.

  — Gratified they are able to address an officer directly.

  — Satisfied the process appears to help hold police and sheriff's departments accountable.

- Police administrators have said that oversight can:

  — Improve their relationship and image with the community.

  — Increase public understanding of the nature of police work.

  — Promote the goals of community policing.

Chapter 1: Introduction

> ## Key Points (continued)
>
> - Improve the quality of the department's internal investigations.
> - Reassure a skeptical public that the department already investigates citizen complaints thoroughly and fairly.
> - Help subject officers feel vindicated.
> - Help discourage misconduct.
> - Improve the department's policies and procedures.
>
> • Elected and appointed officials have indicated that oversight:
>
> - Demonstrates their concern for police conduct to constituents.
> - Can reduce the number, success rates, and award amounts of civil suits against the city or county.
>
> • Members of the community at large have suggested that oversight has helped to:
>
> - Reassure the community that appropriate discipline is being handed out for misconduct.
> - Discourage police misconduct.
> - Increase their understanding of police behavior.
>
> • There are serious limitations to what citizen review can accomplish. To be most effective, citizen oversight must complement other internal and external mechanisms for police accountability.

This chapter explains the purposes of *Citizen Review of Police: Approaches and Implementation* and reports the benefits and limitations that participants attribute to citizen oversight of the police. The report has seven other chapters:

- Chapter 2: nine case studies of citizen oversight.

- Chapter 3: three additional responsibilities oversight systems often undertake: policy and training recommendations, mediation, and early warning systems.

- Chapter 4: recruiting, screening, and training oversight staff.

- Chapter 5: special issues related to citizen oversight, including outreach, structure, openness, and politics.

- Chapter 6: resolving potential conflicts between oversight bodies and police.

- Chapter 7: monitoring, evaluating, and funding oversight systems.

- Chapter 8: organizations, materials, and individuals that can provide assistance with establishing, improving, or evaluating oversight systems.

Following a glossary, appendixes provide sample materials from the jurisdictions studied. In addition to the table of contents, readers may locate specific topics of interest from the key points that precede each chapter and from the index.

What the Publication Is Intended to Do

Audiences and purposes for *Citizen Review of Police*

This report has been written primarily for:

- Local government executives, including mayors and city managers.

- Local legislators, including city council members and county commissioners.

This report will also be of interest to:

- Law enforcement administrators, including chiefs, sheriffs, and their management staff.

- Union leaders.
- Citizen groups and public interest organizations.

Oversight directors may find it helpful to ask new oversight investigators or board members to read the publication to learn more about the field.

Citizen Review of Police describes citizen oversight procedures in nine jurisdictions. The descriptions are intended to:

- Enable jurisdictions that may consider setting up a citizen oversight process to benefit from the experience of communities that have already established procedures.
- Enable jurisdictions that already have citizen review to improve their procedures based on the experiences of these nine cities and counties.

The publication does not promote any particular type of citizen review—or citizen oversight in general. Rather, it is intended to:

- Help jurisdictions decide whether they want to create some form of citizen oversight of police or modify the system they already have.
- Help jurisdictions select a citizen oversight system that will best meet their particular needs.

Citizen Review of Police does not evaluate the nine citizen oversight systems; rather, it describes their operations and the problems they have faced. The report also does not focus on the activities of police and sheriff's department internal affairs units except insofar as they interact with civilian oversight bodies (see "Larger Law Enforcement Agencies Have Internal Affairs Units to Investigate Allegations of Police Misconduct").

The need for the report

There has been a considerable increase in the number of oversight procedures that various cities and counties have implemented in the 1990s (see "A Short History of Citizen Review"). However, many of these procedures have had a troubled history that has involved opposition from concerned citizens and community organizations and from law enforcement agencies and police unions. In many cases, the procedures have been revamped, in some cases litigated, and in at least one city (Washington, D.C.) abandoned.

One reason for controversy in many jurisdictions has been the lack of advance planning for an oversight system.

> The main problem with many citizen review procedures . . . is that they have not had a clear vision of their role and mission . . . . This has usually been the result of a failure of civic leadership. Both community activists and government officials have not taken the trouble to study what other jurisdictions are doing, to borrow the best practices and to learn from their mistakes.[1]

LARGER LAW ENFORCEMENT AGENCIES HAVE INTERNAL AFFAIRS UNITS TO INVESTIGATE ALLEGATIONS OF POLICE MISCONDUCT

Most large police and sheriff's departments have internal affairs (IA) units (sometimes called professional standards units) that investigate allegations of officer misconduct filed by citizens or other officers. In some departments, IA units not only recommend findings to the chief or sheriff, they also recommend the types of discipline (sometimes following guidelines that provide a range of punishments for different types of misconduct).

In some departments, officers' supervisors investigate minor alleged misconduct, leaving serious cases to the IA unit. Some departments use supervisory panels composed of command-level staff who, after reviewing IA's investigation results, come to a finding and, if appropriate, recommend discipline. In smaller departments, the chief or sheriff investigates citizen complaints, or the complaints become a command responsibility.

In all departments, the chief or sheriff makes the final determination of discipline, although in some jurisdictions an appointed or elected official (e.g., police commission) may overrule the decision.

Chapter 1: Introduction

A Short History of Citizen Review

The demand for citizen oversight first occurred in the 1950s and 1960s as a result of the civil rights movement and the perception in many quarters that law enforcement responded to racial unrest with excessive force. Many of these early review procedures were short lived.[1]

Citizen review revived in the early 1970s as urban African-Americans gained more political power and as more white political leaders came to see the need for improved police accountability. Most oversight procedures have come into existence after a high-profile case of alleged police misconduct (usually a shooting or other physical force incident), often involving white officers and minority suspects. Racial or ethnic allegations of discrimination are often at the heart of movements to introduce citizen oversight.[2]

By 2000, citizen review has become more widespread than ever before in the United States. As of early 1998, there were more than 90 citizen review procedures. Almost 80 percent of the largest cities had some form of citizen review.[3] However, only a small fraction of law enforcement agencies in the country had citizen oversight.

---

1. Snow, Robert, "Civilian Oversight: Plus or Minus," *Law and Order* 40 (December 1992): 51–56.

2. Terrill, Richard J., "Civilian Oversight of the Police Complaints Process in the United States: Concerns, Developments, and More Concerns," in *Complaints Against the Police: The Trend to External Review,* ed. Andrew J. Goldsmith, Oxford, England: Clarendon Press, 1991; see also Walker, Samuel, and Vic W. Bumphus, "The Effectiveness of Civilian Review: Observations on Recent Trends and New Issues Regarding the Civilian Review of Police," *American Journal of Police* 11 (4) (1992): 1–26.

3. Walker, Samuel, *Achieving Police Accountability,* Research Brief, Occasional Paper Series no. 3, New York: Center on Crime Communities & Culture, 1998: 5.

[Civilian oversight systems] are often put together quickly and with little thought as to their workability or with much consideration as to how they fit into the review systems already in place.[2]

Citizen Review of Police is intended to make it easier to plan an oversight procedure (or decide how to improve an existing procedure) in a thoughtful manner by presenting the options available for structuring a citizen review mechanism.

Another reason for conflict regarding citizen oversight is that—even with advance planning—public officials, police and sheriff's department executives, union leaders, police officers, and community activists usually have different expectations of what oversight should and can accomplish. This publication should help these parties identify and agree on reasonable and feasible objectives—and dispel unrealistic fears about what the process may do—so they can try to avoid the battles that many other jurisdictions have experienced.

> *Citizen Review of Police is intended to make it easier to plan an oversight procedure in a thoughtful manner by presenting the options available for structuring a citizen review mechanism.*

Features of the Report

Sources of information for the publication

The information presented in this report comes from five principal sources:

1. Literature on citizen oversight of the police (see chapter 8, "Additional Sources of Help").

2. In-person interviews in Berkeley and San Francisco, California; Minneapolis and St. Paul, Minnesota; and Rochester, New York, with oversight staff (directors, board members, auditors, ombudsmen, investigators); complainants; law enforcement administrators, internal affairs investigators, police union leaders, and subject officers; local elected and appointed officials (e.g., city council members, mayors, city managers); and representatives of citizen groups.

3. Telephone interviews with similar individuals in four other communities: Flint, Michigan; Orange County (Orlando), Florida; Portland, Oregon; and Tucson, Arizona.

4. Less comprehensive telephone interviews with other oversight staff across the country (Kansas City, Missouri; Omaha, Nebraska; San Diego and San Jose, California; and Syracuse, New York).

5. Five members of an advisory board assembled to guide and review the publication (see the back of the title page).

The nine jurisdictions studied were selected based on the suggestions of the advisory board. The oversight procedures studied represent a variety of approaches to citizen oversight in different areas of the country and in jurisdictions of varying size and governance (see exhibit 2–1 in chapter 2, "Selected Features of the Nine Oversight Systems").

Terminology used in the report

Different law enforcement agencies use different terminology to denote identical or similar activities. To avoid confusion, *Citizen Review of Police* usually uses the following terms regardless of the local jurisdiction's actual terminology:

- Complainant (sometimes called "appellant").

- Board and board member (sometimes called panelist/panel member, commission/commissioner).

- Executive director or director (sometimes called "officer" or "examiner").

- Police union (also called federation, association).

- Internal affairs (IA) (some departments have renamed their IA units "professional standards").

"Findings That Review Boards and Police Departments Make" identifies and defines the principal terms used to describe possible findings regarding allegations of officer misconduct. A glossary following chapter 8 defines other specialized terms used in the report.

FINDINGS THAT REVIEW BOARDS AND POLICE DEPARTMENTS MAKE

Review boards and police departments generally use a common set of terms to identify the findings that their investigations can lead to:

- Unfounded: The alleged act did not occur, or the subject officer was not involved in the act; therefore the officer is innocent.

- Exonerated: The alleged act did occur, but the officer engaged in no misconduct because the act was lawful, justified, and proper (sometimes called "proper conduct").

- Not sustained: The evidence fails to prove or disprove that the alleged act(s) occurred.

- Sustained: The alleged act occurred and was not justified (e.g., it violated department policy).

Some oversight bodies and police departments come to findings that conclude the subject officer committed an act that was inappropriate but that hold the *department* responsible for the officer's misconduct:

- Policy failure: Department policy or procedures require or prohibit the act (e.g., an officer may not use a cruiser to drive someone to a bus stop whose car was towed).

- Supervision failure: Inadequate supervision— the officer's sergeant or lieutenant should have informed the officer not to engage in the act or to discontinue it (e.g., a sergeant asks a supervisor, "Here's what I've got. Is that probable cause to arrest the guy?" and the supervisor gives the officer bad advice).

- Training failure: The officer receives inappropriate or no training in how to perform the act properly (e.g., distinguishing an intoxicated person from someone going into diabetic shock).

Types of Citizen Review

According to experts, "There is no single model [of citizen oversight], and it is difficult to find two oversight agencies that are identical."[3] However, most oversight systems fall into one of four types:[4]

- Type 1: *Citizens investigate allegations* of police misconduct and *recommend findings* to the chief or sheriff.

- Type 2: Police officers investigate allegations and develop findings; *citizens review and recommend* that the chief or sheriff approve or reject the findings.

- Type 3: Complainants may *appeal findings* established by the police or sheriff's department *to citizens*, who review them and then recommend their own findings to the chief or sheriff.

- Type 4: An auditor *investigates the process* by which the police or sheriff's department accepts and investigates complaints and reports on the thoroughness and fairness of the process to the department and the public.

While some oversight procedures represent "pure" examples of these models, many oversight systems are hybrid models that merge features from the four different types into their own unique formulation. For example, the Office of Community Ombudsman in Boise, Idaho, created in 1999, combines the authority to investigate complaints—a type 1 oversight system—with the responsibility to review internal affairs investigations to determine if they are thorough and fair—a type 4 oversight system.[5]

Furthermore, as discussed in chapter 3, "Other Oversight Responsibilities," any oversight system may undertake three other responsibilities in addition to investigating, reviewing, or auditing citizen complaints:

1. Recommending changes to department policies and procedures and suggesting improvements in training.

2. Arranging for informal or formal mediation.

3. Assisting the police or sheriff's department to develop or maintain an early warning system for identifying potentially problematic officers.

To make an informed decision about which type of oversight procedure to adopt and which additional responsibilities to undertake, jurisdictions need to examine tradeoffs inherent in fashioning an oversight system—what they will gain and lose by the approach they select. Only with these tradeoffs in mind can communities select a system that will best meet their local needs, resources, and constraints. Exhibit 1–1 lists some of the tradeoffs jurisdictions need to consider in selecting an oversight procedure.

In addition to weighing tradeoffs, selecting oversight features may depend on several criteria:

> *To make an informed decision about which type of oversight procedure to adopt and which additional responsibilities to undertake, jurisdictions need to examine tradeoffs inherent in fashioning an oversight system—what they will gain and lose by the approach they select.*

- Which features does the public want?

- Which features are most effective in achieving the goals the community expects the oversight procedure to achieve?

- Which features may create conflict with the police or sheriff's department or the police union, and which features may disappoint community activists?

- How much will the features cost?

- How will the new features mesh with existing oversight procedures?

Potential Benefits of Citizen Oversight

Oversight systems have the potential to benefit complainants, police and sheriff's departments, elected and appointed officials, and the public at large. The extent to which benefits materialize depends not only on the type of oversight procedure implemented but also, and critically, on how well these groups work together. The working relationships among the groups in turn depend to a tremendous extent on the personality, talents, dedication,

Exhibit 1–1. Sample Tradeoffs Jurisdictions Need to Consider in Choosing an Oversight Procedure

- *Volunteers* versus *paid staff.* Volunteer participants are lay community members who represent the concerns of the public. Professionals conduct the day-to-day work of citizen oversight, carrying out the public's wishes. On one hand, an oversight procedure involving only paid staff usually will not be as representative of the community as will a system that uses volunteers. On the other hand, the amount of time required to provide adequate oversight can normally be provided only by one or more paid staff who have been hired specifically to dedicate themselves to oversight activities. As a result, many oversight procedures use volunteers and paid staff.

- *Public hearings* versus *private hearings.* Public hearings may make the community feel it has more control over police misconduct because officers' alleged misconduct is made known. Private hearings are simpler logistically and protect complainants and officers from public exposure.

- *Investigative authority* versus *review authority.* Investigating complaints can help ensure they are done thoroughly and fairly, but hiring investigators can be expensive. Reviewing cases is less expensive but requires department cooperation in sharing records. Some oversight systems have both responsibilities.

- *Taking on additional responsibilities*—policy recommendations, mediation, or early warning systems:
  — Developing policy recommendations may involve a conflict of interest because investigating and reviewing cases requires impartiality, but developing policy recommendations may involve political advocacy. However, providing policy recommendations may expand the oversight body's influence.
  — Mediation, usually held in private and kept confidential, may have less "teeth" than a public hearing. However, mediation may encourage citizens to file complaints; save the time and expense of a hearing; and educate officers about the impact of their words, behaviors, and attitudes on the public.
  — Early warning systems can help identify potentially troublesome officers and may deter officer misconduct, but they may alienate officers if unsustained cases are included.

- Accepting complaints *directly* versus accepting them only *by referral from the police or sheriff's department.* Citizens who may be reluctant to file complaints with the department may file with the oversight body, but outreach must be conducted to make citizens aware of this option.

flexibility, and open-mindedness of the principal actors in each group—in particular, the oversight director, the chief of police or sheriff, union leaders, the mayor, city council members, and the city manager.

Potential benefits to complainants

Citizen oversight can have three benefits for complainants. Oversight can:

1. Help complainants feel "validated" in the minority of instances in which oversight bodies agree with their allegations.

   > I was afraid the investigation would be rush-rush, but it was very thorough. Before the hearing, the investigator was very comforting toward my son, who was only 16 years old, going over the process in detail with him. When I received a letter after the hearing that my son's allegations had been sustained, I was surprised. I didn't have faith in the powers that be to be objective. I was elated that my son had been heard and that the officer had to sit through the entire hearing. My son was happy, too; he didn't think he'd win either.
   > —mother of a juvenile complainant

   > The phenomenon of complainants [who] feel validated because the oversight body agrees with their allegations is only part of the story. As the procedural justice literature suggests, the process is as important as the outcome. People feel validated when they feel they have an opportunity to be heard. Civilian oversight is likely to enhance that feeling by virtue of appearing to be independent of the police department.
   > —Samuel Walker, Professor, University of Nebraska at Omaha

2. Give complainants the satisfaction of expressing their concern in person to the officer when oversight includes a mediation option.

> Many complainants just want to be able to express their anger or concern face to face with the officer in an impartial setting without being cut off, and that is all they need.
> —Jackie DeBose, Berkeley Police Review Commission board member

3. Help hold the police or sheriff's department accountable for officers' behavior.

   > I felt I had done my civic duty. This was a young cop [I complained about]. I coach people all the time [at his job], so I wanted this officer to get better supervision and training so that in a similar event he would not engage in the same misconduct. I felt good; the officer got the direction he needed.
   > —a complainant

   > [R]eview [by the Police Review Commission] of this incident [in which the commission exonerated officers of a complaint that their use of excessive force resulted in a man's death] prompted development of a new *Berkeley Police Department Training and Information Bulletin* regarding the risk of asphyxiation during four-point restraints. Development of this bulletin was a pivotal issue in bringing closure for the family and ensuring that their tragedy had some positive effect.
   > —Robert Bailey, former assistant city manager, Berkeley

Potential benefits to police and sheriff's departments

As summarized in exhibit 1–2, police and sheriff's department personnel have identified several possible benefits citizen oversight can provide them, depending on the type of oversight procedure adopted. Oversight can:

1. Improve the department's relationships and image with the community by:

   - Helping to establish and maintain the department's reputation for fairness and firmness in addressing allegations of police misconduct.

   > The board takes a lot of pressure or criticism off IA and the chief because citizens are making the decisions about misconduct and the department can't be accused of a coverup.
   > —an IA commander

EXHIBIT 1–2. POTENTIAL BENEFITS OF CITIZEN OVERSIGHT FOR POLICE AND SHERIFF'S DEPARTMENTS

Law enforcement managers and line officers report that citizen oversight can provide a number of benefits to police and sheriff's departments depending on the type of oversight procedure adopted:

1. Improve the department's relationship and image with the community by:
    a. Helping to establish and maintain its reputation for investigating alleged officer misconduct with fairness and firmness.
    b. Helping to reduce community concerns about possible police coverups in high-profile cases.

2. Increase the public's understanding of police work, including the use of force.
3. Promote the goals of community policing.
4. Improve the quality of the department's internal investigations of alleged misconduct.
5. Reassure the public that the department's internal investigations of citizen complaints and its process for disciplining officers already are thorough and fair.
6. Help subject officers feel vindicated.
7. Help discourage misconduct among some officers.
8. Improve department policies and procedures.

Oversight makes the department's job easier because, if we couldn't point to the board's sustained rate of 10 percent, we would be criticized and accused of a coverup [because our internal rate would be just as low].
—a deputy chief

- Helping reduce or eliminate community concerns regarding specific high-profile incidents of alleged misconduct.

Two Rochester police officers arrested two individuals on drug-dealing charges. The mother of one of them claimed the two youths had been innocently walking down the street when the officers approached them. One officer got into a tussle with the mother's son and, the mother said, threw her son through a store window. Some members of the community were outraged at what they felt was police brutality. When the Civilian Review Board [CRB] heard the case, it learned that the two men had drugs in their possession. In addition, the store owner testified that the officers had bent over backwards to be polite to the men—and that the son had pushed the officer into the store window. Because the CRB exonerated the officers, the community calmed down.
—Andrew Thomas, Executive Director, Rochester Center for Dispute Settlement

> Community policing is related to citizen review. It's another way to communicate with the public, another source of community input.

We love being able to send cases to the board because we get less pressure from liberal groups about not properly disciplining officers.
—an IA commander

2. Increase public understanding about the nature of police work, such as the occasions when officers need to use force. Help the public develop realistic expectations regarding actions officers are allowed to take—or departments have the personnel to take—to abate crime and disorder.

3. Promote the goals of community policing. According to the Berkeley Police Review Commission 1996 annual report:

Community Involved Policing, especially its "Problem Solving" method of organizing police work, depends heavily on the involvement of especially those citizens who are demographically and geographically closest to crime and criminals. Therefore, it is undermined by hostility generated in the normal unfolding of police/citizen interactions at precisely the point at which it needs the most support.

Some police administrators agree.

Community policing is related to citizen review. It's another way to communicate with the public, another source of community input.
—Fred Lau, Chief, San Francisco Police Department

ABOLISHING CITIZEN OVERSIGHT WILL NOT SAVE A POLICE OR SHERIFF'S DEPARTMENT MONEY

Getting rid of the oversight body will not save the police or sheriff's department money. When the Minneapolis City Council was considering abolishing the Citizen Police Review Authority (CRA), Lt. Robert Skomra, the IA commander at the time, examined the number of cases CRA handled. Skomra determined that, if CRA disappeared, the police department would have to find the funds to at least double and possibly triple the number of existing IA investigators. The department would also have had to find desk space for the new investigators. Minneapolis police Chief Robert Olson agreed: "If the CRA were abolished, I would have to hire additional IA investigators."

Chapter 1: Introduction

For community policing to be effective, the integrity of the agency and the community's trust in it are critical. The CRB [Citizen Review Board] contributes to the trust because nonsworn citizens are involved in reviewing the agency.
—Kevin Beary, Orange County (Florida) Sheriff

4. Improve the quality of the department's internal investigations of alleged misconduct.

The board has improved our professional standards' investigative reports because investigators get dressed down and embarrassed at hearings for any sloppiness, such as drawing conclusions on flimsy evidence. As a result, if there is any litigation on the complaint, the report will enhance the agency's position.
—Capt. Melvin Sears, Orange County Sheriff's Office administrative coordinator to the Citizen Review Board

Informally in discussions after hearings and in the questions board members ask of PSD [professional standards division] investigators during hearings, [board] members have made observations about deficiencies in the investigators' reports that have resulted in improved reporting. For example, board members kept objecting to the way officers and investigators included opinions in their reports, rather than just the facts.
—Maj. Karon LaForte, Orange County Sheriff's Office IA commander

Investigators do a better job investigating cases because they know that PIIAC [Police Internal Investigations Auditing Committee] will be looking at their work product, so they are less likely to take shortcuts in their research and reporting than in the past.
—Charles Moose, former Chief, Portland Police Bureau

> *Investigators do a better job investigating cases because they know that PIIAC will be looking at their work product.*
> —Charles Moose, former Chief, Portland Police Bureau

5. Help reassure the public that the department already investigates citizen complaints thoroughly and fairly.

Even when the department is capable of imposing appropriate discipline without citizen review, an oversight procedure can reassure skeptical citizens that the agency is doing its job in this respect.
—Douglas Perez, former Deputy Sheriff, Professor, Plattsburgh (New York) State University

6. Help some subject officers feel vindicated. The St. Paul oversight board exonerated an officer after a citizen complained about an allegedly offensive remark the officer had made to a block party. When the officer and Donald Luna, the board chair, happened to meet at a graduation ceremony, the officer said:

I want to thank you for the letter to the chief. I'd put in a lot of time and felt I had deescalated a tense situation. I couldn't believe there had been a complaint; I felt I deserved an award. I felt the commission understood me.

7. Help discourage misconduct (see below).

If I live a normal lifespan, I'm a citizen longer than I'm a cop, so I want a system of checks and balances to help prevent police misconduct.
—Trevor Hampton, former Chief, Flint Police Department

8. Improve the department's policies and procedures (see chapter 3, "Other Oversight Responsibilities").

Potential benefits to elected and appointed officials

By establishing or improving a citizen oversight mechanism, local officials can demonstrate their concern to eliminate police misconduct—or publicize a department's existing exemplary police behavior. Officials may also be able to reduce the number of civil lawsuits (or successful suits) against the city or county or the dollar value of successful awards. These suits can be expensive.[6] During

SOME POLICE CHIEFS HAVE ESTABLISHED CITIZEN OVERSIGHT PROCEDURES ON THEIR OWN

Police chiefs have taken the initiative to establish citizen oversight procedures on their own.

- When Robert Olson, chief of the Minneapolis Police Department since 1995, was commissioner in Yonkers, New York, he established a civilian oversight program because of the department's poor relations with the African-American community. The review board he established included four civilians nominated by a citizen panel, an officer nominated by the police union, and three other officers Olson selected. He approved all the candidates. The board, which met monthly, reviewed completed IA cases and occasionally pending cases, and it had the authority to direct IA to conduct additional investigations. The board recommended findings, with Olson retaining the ultimate decision to decide cases and impose discipline.

- William Finney, chief of the St. Paul Police Department, recommended the Police Civilian Internal Affairs Review Commission on his own initiative because he felt the need to gain citizens' perspective on department behavior. (See the St. Paul case study in chapter 2.)

a 6-month period alone, deaths and injuries resulting from police shootings resulted in more than 300 civil suits against the Washington, D.C., police department, with nearly $8 million in court settlements and judgments awarded.[7]

- Joan Campbell, the chairperson of the Minneapolis City Council Ways and Means Committee, reports that, when citizens sue the city for alleged police brutality, the judge asks if the Civilian Police Review Authority (CRA) sustained the case. In many instances in which CRA has not, the council has a stronger case for not settling with the complainant and for expecting the judge to rule in the city's favor. As a result, the city has gone to court on more cases and won most of them. Campbell also believes that CRA has reduced the number of complaints that have gone to litigation because complainants feel they have already had their day in court with the review board.

- According to Robert Bailey, former assistant city manager in Berkeley, the Police Review Commission "saved the city at least $100,000 from one potential lawsuit alone." Because they did not trust the police to investigate the matter fairly, family members filed a complaint after a relative died from cardiac arrest in police custody after being put into four-point restraints. The board decided to hear the case en bloc and hired an independent toxicologist to review the medical records and do more testing. The toxicologist, as had the coroner previously, reported that use of force had not caused the person's death—aspiration due to a drug overdose was the cause. The family decided not to sue the city after the board concluded that the officers did not use excessive force.

> *According to Robert Bailey, former assistant city manager in Berkeley, the Police Review Commission "saved the city at least $100,000 from one potential lawsuit alone."*

- Merrick Bobb, special counsel to Los Angeles County, reported: "In 1992 . . . the County of Los Angeles had 800 police misconduct cases pending. And the exposure to the taxpayers of the County of Los Angeles was calculated by the County's lawyers as far in excess of $600 million. Today, 5 years later, in 1997, we find the caseload has dropped from 800 cases to a little over 200 cases. We find that the amount of money that is being spent has dropped for the first time to below the 10 million mark in terms of judgments, settlements, and attorneys' fees in such cases. . . . I think this is a testament to the effect of civilian oversight, civilian review" initiated in 1993.[8]

Chapter 1: Introduction

- Troubled in part by fatal shootings by Albuquerque police officers (31 in 10 years) and extremely high annual payments for tort claims involving police officers (up to $2.5 million per year), the Albuquerque city council hired two consultants in 1996 to evaluate the city's existing oversight system and recommend alternatives.[9]

Potential benefits to the community at large

Citizen oversight can benefit the entire community, not just individual complainants. Oversight can:

1. Help to reassure the community that appropriate discipline is being imposed. Even when departments are doing a top-notch job disciplining errant officers, the public may lack confidence in the process. An oversight procedure that provides citizens with a window into how the department operates can change the opinion of these concerned citizens.

2. Help discourage police misconduct. While there is no empirical evidence that oversight bodies can deter police misconduct,[10] there are three ways in which citizen review may help encourage officers to act appropriately.

   - When oversight bodies recommend that an officer be retrained, the officer may learn how to avoid the type of behavior that led to the citizen complaint.

   - When police and sheriff's departments adopt policy and procedure changes that oversight bodies recommend, officers may have a better understanding regarding how they should perform their job.

   - Oversight bodies may discourage some officers from engaging in misconduct by reducing their chances for promotion.

   > I was nervous about whether a sustained case might hamper my promotion to lieutenant. The chief had made it plain that an officer with sustained complaints would not be looked at as favorably for promotion as officers with no or fewer complaints. If you look at the people he's passed over, you can see that the officers with complaints have been passed over.
   > —a lieutenant

   > The [review] board influences assignments to [desirable] details. We have supervisors in units now who don't want "cowboys" in their units, so officers with complaints could get passed over.
   > —an officer

3. Increase public understanding of police policies, procedures, and behavior. Complainants learn about police procedures from oversight investigators, board members, and officers during mediation. Board members themselves become better educated about police procedures and can share their understanding with other members of the community.

   Finally, by holding special public hearings, oversight bodies may be able to defuse tense community conflicts, channeling anger into constructive solutions. Berkeley's charter requires the Police Review Commission (PRC) to hold hearings at the request of board members or voters. PRC held a special public hearing after University of California police officers were accused of using excessive force against students during a campus demonstration. Although contentious, the meeting resulted in recommendations regarding officer conduct during demonstrations—several of which the department implemented—to help prevent future discord.

> *Citizen oversight systems need to be part of a larger structure of internal and external police accountability.*

Limitations to Citizen Review

As summarized in exhibit 1–3, citizen oversight has several inherent and potential limitations.

Citizen oversight systems need to be part of a larger structure of internal and external police accountability; by itself, citizen oversight cannot ensure that police will act responsibly. An evaluation of New York City's oversight system concluded, "In general, civilian complaint review procedures appear to be a necessary but insufficient component of the [New York City Police] Department's approach to controlling officer misconduct."[11] "Supplements

Exhibit 1–3. Limitations to Citizen Oversight

1. Citizen oversight cannot by itself ensure police accountability. Jurisdictions need to implement other internal and external mechanisms to achieve this goal.

2. The effectiveness of citizen oversight depends enormously on the talent, fairness, and personalities of the principal individuals involved.

3. Oversight bodies have limited authority; they do not impose discipline or dictate department policies or procedures.

4. The findings some oversight bodies make, or the investigations they conduct, have no influence on some police managers.

5. Oversight bodies typically fail to hold department supervisors responsible for line officers' behavior.

6. Some complainants who lose their cases express disappointment with the oversight process.

7. When long delays occur between filing a complaint and its resolution, complainants become frustrated and disillusioned—even when they win the case.

8. Some complainants and a small minority of other individuals will not be satisfied with the actions of police officers and deputy sheriffs no matter what the oversight body does.

9. Oversight procedures in some jurisdictions have exacerbated tensions among local officials, police and sheriff's departments and unions, and citizen groups and activists.

to Citizen Oversight" suggests other procedures that, taken together with citizen oversight and an effective internal affairs unit, may improve police accountability in departments in which officer conduct needs improvement.

The effectiveness of citizen oversight depends enormously on who the principal parties are. In Minneapolis, there was "a complete turnaround" in the relationship between the police department and the Civilian Police Review Authority after a new chief and a new executive director took over and the new mayor made clear she expected them to cooperate. Supporting this observation, a subject officer in Minneapolis wrote on his anonymous customer satisfaction survey in 1998, "It appears as though there have been some changes in the factfinding process, which resulted in a more satisfactory outcome. In the past I was unable to give a favorable opinion of the Civilian Police Review Authority, but I was pleased with this openness."

We need to review higher-ups' behavior to produce accountability among line officers. Otherwise, the beat officer gets scrutinized and the supervisors are never held accountable, never called to account.

—Mary Dunlap, director of San Francisco's Office of Citizen Complaints

Oversight bodies in the United States have limited authority. In particular, they do not have the power to discipline officers or establish department policies. In these areas, they are only advisory. Furthermore, oversight bodies have no influence on some police managers or, as a result, many or most line officers. According to one chief, "Boards can't be effective because officers fear IA, not them." Concerns about liability and supervisor criticisms may typically discourage misconduct much more than either citizen oversight or internal affairs investigations.

In a related vein, oversight procedures generally focus on individual officers, letting supervisors off the hook in terms of management's responsibility for—and tremendous influence over—line officers' and deputies' behavior. As a result, unless the oversight system includes making recommendations for policy and procedure changes and has the ability to influence their adoption, department supervisory and training practices that may be allowing misconduct to occur will

SUPPLEMENTS TO CITIZEN OVERSIGHT*

Several other procedures for maintaining citizen oversight of law enforcement agencies can supplement a citizen review process. One or more of the alternatives listed below can also *substitute* for citizen review in certain cases of alleged police misconduct.

Legislative Control

Legislatures can monitor police behavior through investigations, appropriations pressure, oversight committees, and other means.

Civil Litigation

Complainants may sue police officers in State court and seek common law tort remedies. They may also sue in Federal court for violations of Federal civil rights.

Criminal Prosecution

Prosecutors at the local, State, and Federal levels can apply applicable criminal statutes to situations involving alleged police misconduct.

Federal Government Suits

Under a 1994 law, prosecutors may seek changes in the operations of local police departments in Federal courts. Suits by the U.S. Department of Justice can require reform through court-approved agreements in which police departments agree to change the way they track and handle citizen complaints and disciplinary decisions or by installing a Federal monitor to oversee the department's activities in these areas.

Supervisor Accountability

There are several internal actions police and sheriff's departments can take, if needed, that may make a significant difference in helping to prevent police misconduct, including effective applicant screening, recruit and inservice training, peer review, and, perhaps most important, leadership training. Lt. Bret Lindback with the Minneapolis Police Department emphasizes that chiefs and sheriffs should be given funding to provide:

> the best leadership training you can find to make supervisors and managers accountable for what their guys do on the street.... You need to train them to tell line officers, "You don't do that [misconduct, discourtesy] on my watch." A week's training when you get your sergeant's bar isn't enough. You need ongoing training, two or four times a year, to build good leadership skills.

Mary Dunlap, director of San Francisco's Office of Citizen Complaints (OCC), agrees:

> We need to review higher-ups' behavior to produce accountability among line officers. Otherwise, the beat officer gets scrutinized and the supervisors are never held accountable, never called to account.

---

* For a more complete discussion of the alternatives, see Perez, Douglas W., *Common Sense About Police Review*, Philadelphia: Temple University Press, 1994: 48–63.

SUPPLEMENTS TO CITIZEN OVERSIGHT (CONTINUED)

OCC addresses supervisors who share responsibility for officers' misconduct by charging them with failure to supervise the accused officer properly. In *Banta v. City and County of San Francisco* (1998), the presiding judge of the Superior Court dismissed a challenge to OCC's power to add an allegation against a sergeant for failure to supervise.

In the last analysis, supervisor accountability extends to the chief or sheriff, who must exercise active responsibility for ensuring that his or her officers and deputies comport themselves appropriately. If the chief executive will not or cannot ensure proper conduct, it is the obligation of the mayor, city manager, or city council to find a new chief and the duty of the voters to elect a new sheriff.

---

remain untouched. As one commentator observed, "The solution to rotten apples is to fix the police barrel."[12] Some police chiefs and sheriffs agree that they should be held accountable for preventing misconduct, and, if they fail, they should be dismissed. One chief commented, "If IA is not up to snuff, give the chief a chance to fix it, and, if he doesn't, fire him. So the solution [to police misconduct] is to hold the chief accountable."

Some complainants who lose their cases (and even some who win) feel dissatisfied with the process, the results, or both. Others are frustrated that they cannot find out what the chief's or sheriff's finding was or whether and what kind of discipline was imposed. According to Jackie DeBose, a member of Berkeley's board, "I have run into several citizens who lost their cases, and they were livid—they felt they had been done an injustice." The Vera Institute of Justice in New York City surveyed a sample of 371 citizens who had filed complaints with the city's Citizen Complaints Review Board.[13] The Vera Institute concluded that "the investigative process itself has a significant negative influence" on citizen satisfaction because of how long the process took and the lack of contact with and information about the subject officer and the final outcome. Some complainants, and a small minority of the public, will not be satisfied with any actions oversight bodies take. These individuals may have unreasonable expectations of how the police should behave or unreasonable hopes for what citizen oversight procedures can accomplish.

Finally, oversight procedures in some jurisdictions have exacerbated tensions among local officials, police and sheriff's departments and unions, and citizen groups and activists. This worsening of the status quo has occurred for many reasons, such as unrealistic expectations on the part of activists or unrealistic apprehensions by police and sheriff's departments about what the oversight procedure would accomplish; failure to involve all affected parties in the planning process; biased oversight staff; inadequate funding leading to long delays in case processing; and political motives for setting up the procedure on the part of local officials.

Despite these limitations, local government officials, law enforcement managers, and citizens in many jurisdictions believe that citizen oversight can be of value. The following chapters illustrate the potential benefits of citizen review as well as its limitations.

Notes

1. Walker, Samuel, *Achieving Police Accountability*, Research Brief, Occasional Paper Series, no. 3, New York: Center on Crime Communities & Culture, 1998: 5.

2. Snow, Robert, "Civilian Oversight: Plus or Minus," *Law and Order* 40 (December 1992): 51–56.

3. Luna, Eileen, and Samuel Walker, "A Report on the Oversight Mechanisms of the Albuquerque Police Department," prepared for the Albuquerque City Council, 1997: 121.

4. Walker, Samuel, *Citizen Review Resource Manual*, Washington, D.C.: Police Executive Research Forum, 1995.

CHAPTER 1: INTRODUCTION

5. An example of an oversight procedure that does not fall into any of these four types exists in Charlotte, North Carolina. The city's Community Relations Committee appoints a staff member to attend the Charlotte-Mecklenburg Police Department's internal hearings of serious allegations against officers. The staff member contributes to the findings of each review panel and can give a minority report to the chief, city manager, and city council (which happened once).

6. "How Much Force Is Enough?" *Law Enforcement News* 24 (500) (November 30, 1998): 1. "The cost of a civil suit goes beyond expenses incurred by individual police officers. Such factors as the cost of liability insurance, litigation expenses, out-of-court settlements, and punitive damage awards all make civil liability an extremely expensive proposition for police officers, law enforcement agencies, governments, and, ultimately, taxpayers. . . . After several lawsuits are filed, . . . premium prices can skyrocket, or companies may refuse to ensure the department." Gaines, Larry K., Victor E. Kappeler, and Joseph B. Vaughn, *Policing in America,* 2d ed., Cincinnati: Anderson Publishing Company, 1997: 294.

7. "How Much Force Is Enough?", 1.

8. National Association for Civilian Oversight of Law Enforcement conference transcript, October 15–17, 1997, Lanham, Maryland.

9. Luna and Walker, "A Report on the Oversight Mechanisms of the Albuquerque Police Department."

10. The oversight process lacks two qualities thought to be essential to deter misconduct: certainty and severity of punishment (Sviridoff, Michele, and Jerome E. McElroy, *Processing Complaints Against Police: The [New York City] Civilian Complaint Review Board,* New York: Vera Institute of Justice, 1988: 35). Oversight procedures also often lack a third critical element for deterrence: swiftness.

11. Sviridoff and McElroy, *Processing Complaints Against Police,* 38.

12. Bayley, David, "Getting Serious about Police Brutality," in *Accountability for Criminal Justice: Selected Essays,* ed. Philip C. Sternberg, Toronto: University of Toronto Press, 1995: 96.

13. Sviridoff and McElroy, *Processing Complaints Against Police.*

Chapter 2: Case Studies of Nine Oversight Procedures

This chapter presents brief case studies of nine oversight procedures arranged alphabetically by jurisdiction. The case studies concentrate primarily on the operational procedures of the oversight mechanisms. Details about other aspects of the jurisdictions' procedures are presented in other chapters of the report:

- Chapter 3, "Other Oversight Responsibilities," describes how the jurisdictions develop policy and procedure recommendations, implement mediation, and assist with early warning systems.

- Oversight staffing arrangements are discussed in detail in chapter 4.

- Chapter 5, "Addressing Important Issues in Citizen Oversight," presents such problematic areas as intake, outreach, and "politics."

- Chapter 6 identifies the most common areas of conflict between oversight mechanisms and police and sheriff's departments.

- Monitoring, evaluation, and funding issues are addressed in chapter 7.

Exhibit 2–1 identifies the location, type of system, principal activities, and paid staff and budget for each oversight mechanism. Exhibit 2–2 summarizes the number of complaints, hearings, mediations, and other pertinent activities each system conducted in 1997, the extent to which its proceedings are open to the public, whether it has subpoena power, and the types of complaints it reviews.

As the exhibits illustrate—and the case studies that follow explain—there is enormous variation in the structure and operations of the nine systems. In fact, dissimilarity, rather than similarity, is the rule among the nine systems. In part this is because radically different systems were selected for inclusion in this report to illustrate the diversity in oversight mechanisms from which other jurisdictions can choose if they wish to develop a procedure of their own or modify an existing one. However, the diversity also reflects the fact that local officials have shaped their oversight systems very differently to accommodate unique local pressures (e.g., from activist groups, police unions, or office holders), legal considerations (e.g., with regard to the types of information that can be made public or the provisions of labor-management agreements), funding resources, and honest disagreements about what would work best in their communities.

All four types of oversight approaches listed in footnote a in exhibit 2–1 are represented among the nine oversight systems. However, two jurisdictions have combined two different approaches: Portland has a citizen appeals board (type 3) and an auditor who monitors the department's complaint investigation process (type 4), while Tucson has a citizen board that reviews internal affairs findings (type 2) and also an auditor (type 4). Other "models" are not pure either; for example, while San Francisco's Office of Citizen Complaints involves citizens in investigating complaints (type 1), OCC staff also prosecute cases at chief's hearings and before the police commission, a responsibility—and expense—that goes well beyond that of investigating complaints. Similarly, the Minneapolis Civilian Police Review Authority (CRA) not only hires professional staff to investigate complaints (type 1) but its volunteer board members also hold hearings for complaints for which investigators have found probable cause. Furthermore, the CRA executive director prosecutes these cases before the civilian review board. While San Francisco's OCC and Minneapolis' CRA both investigate most complaints in place of internal affairs, Berkeley's Police Review Commission investigates cases simultaneously with internal affairs investigations. The St. Paul Police Civilian Internal Affairs Review Commission and San Francisco's Office of Citizen Complaints recommend discipline to the chief.

The independence of the nine oversight systems also varies considerably. The St. Paul police chief proposed

CHAPTER 2: CASE STUDIES OF NINE OVERSIGHT PROCEDURES

EXHIBIT 2–1. SELECTED FEATURES OF THE NINE OVERSIGHT SYSTEMS

Location	Name	Type[a]	Principal Activities	Paid Staff	Budget
Berkeley pop.: 107,800 sworn: 190	Police Review Commission	1	• board hears complaints in public hearings and recommends findings to the city manager • board and IA investigate many cases simultaneously • board recommends policy changes	4 full time	$277,255
Flint pop.: 134,881 sworn: 333	Office of the Ombudsman	1	• office investigates complaints against all city departments • office reports subject officers' names to the media • chief relies primarily on internal affairs' (IA's) investigation results	2 full time[b] 1 part time	$173,811
Minneapolis pop.: 358,785 sworn: 919	Civilian Police Review Authority	1	• paid professionals, not IA, investigate complaints • volunteer board members hear valid complaints with probable cause • half of cases with probable cause are stipulated • many cases are professionally mediated	7 full time	$504,213
Orange County pop.: 749,631 sworn: 1,134	Citizen Review Board	2	• board members review IA findings on use-of-force cases • board members make policy recommendations • a sheriff's captain coordinates the board's activities	2 part time	$20,000
Portland pop.: 480,824 sworn: 1,004	Police Internal Investigations Auditing Committee	3, 4	• citizens hear appeals; the city council hears further appeals • citizen advisers review closed cases • an examiner audits IA investigations • all three provide policy recommendations	1 full time	$43,000
Rochester pop.: 221,594 sworn: 685	Civilian Review Board	2	• board members are trained mediators • specially trained board members conduct conciliations • board provides policy and training recommendations	1 full time 3 part time	$128,069
St. Paul pop.: 259,606 sworn: 581	Police Civilian Internal Affairs Review Commission	2	• police chief established board as inhouse review procedure • board, not IA, recommends discipline • seven-member board includes two St. Paul police officers	1 part time	$37,160
San Francisco pop.: 735,315 sworn: 2,100	Office of Citizen Complaints	1	• office, not IA, investigates most complaints • office prosecutes cases in chief's hearings and police commission trials • office recommends policy and training changes	30 full time	$2,198,778
Tucson pop.: 449,002 sworn: 865	Independent Police Auditor and Citizen Police Advisory Review Board	2, 4	• auditor reviews completed cases • auditor sits in on ongoing cases and asks questions • board can review completed cases • board hears community's concerns about the police • auditor and board make policy recommendations	2 full time	$144,150

a. Type 1: citizens investigate allegations and recommend findings; type 2: police officers investigate allegations and develop findings; citizens review findings; type 3: complainants appeal police findings to citizens; type 4: an auditor investigates the police or sheriff's department's investigation process.
b. This represents the proportion of staff who handle complaints against the police.

Exhibit 2-2. Additional Features of the Nine Oversight Systems

System	Activities*	Openness to Public Scrutiny	Subpoena Power	Complaints Reviewed
Berkeley Police Review Commission (PRC)	42 complaints filed and investigated 12 hearings	• hearings and commission decisions are open to public and media • general PRC meetings available for public to express concerns • appeal process • IA's dispositions and discipline not public	yes	• all types of complaints
Office of the Flint Ombudsman	313 cases investigated	• findings distributed to media and city archives • no appeal • chief's finding public, but not discipline	yes, but never used	• all types of complaints
Minneapolis Civilian Police Review Authority (CRA)	159 formal complaints investigated 14 cases mediated 6 stipulations 3 hearings	• hearings are private • general public invited to monthly CRA meeting to express concerns • appeal process • complainant told whether complaint was sustained • chief's discipline not public until final disposition	no, but testimony required under *Garrity* ruling	• all types of complaints except conduct for which officers can be fired or charged criminally
Orange County Citizen Review Board	45 board hearings	• hearings are open to public and media scrutiny • findings and the sheriff's discipline are matters of public record • no appeal	yes, but never used	• allegations of use of excessive force • discharges of firearms • abuse of power
Portland Police Internal Investigations Auditing Committee (PIIAC)	21 appeals processed 98 audits of completed cases	• PIIAC audits are open to public and media • citizen advisory subcommittee meetings are open to public and media • appeal to city council • PIIAC decisions are public; chief's discipline is not	yes	• all types of complaints
Rochester Civilian Review Board	26 cases reviewed 4 cases mediated	• reviews are closed • results are not public • no appeal	no	• allegations of use of excessive force • conduct that, if proven, would constitute a crime • other matters as determined by the chief
St. Paul Police Civilian Internal Affairs Review Commission	71 cases reviewed	• hearings are closed • no appeal • no publicizing of disciplinary recommendations	yes, but never used	• allegations of use of excessive force • discharge of firearms • discrimination • poor public relations • other matters as determined by the chief or mayor
San Francisco Office of Citizen Complaints	983 cases investigated and closed 50 chief's hearings 6 police commission trials	• chief's hearings are closed • police commission hearings are generally public • appeal process for officers • complaint histories and findings are confidential • chief's discipline not public	yes	• all types of complaints
Tucson Independent Police Auditor and Citizen Police Advisory Review Board	96 complaints 63 investigations monitored (9/1/97 to 6/30/98)	• monitoring is private • appeal process • board holds monthly meeting for public to express concerns	no	• all types of complaints

* Data from 1997 unless indicated otherwise.

the idea of a review commission, nominates its two sworn members, funds it, and houses it in the public safety building. While San Francisco's OCC operates independently of the police department and is funded by the city, its budget is a line item within the police department budget. The OCC director reports to the San Francisco Police Commission, which tries major discipline cases and ratifies policy and training changes. By contrast, an independent dispute resolution center operates Rochester's Civilian Review Board, appoints board members, and receives funding to staff the procedure directly from the city council.

Oversight systems differ in other respects. While most systems may provide policy and procedure recommendations to the local police or sheriff's department, San Francisco's charter requires it to make policy recommendations. In addition, some systems rarely make recommendations, while others are constantly proposing them. Minneapolis and Rochester make considerable use of mediation. Until 1999, Berkeley's mediation system, although required by statute, was dormant, while San Francisco has had difficulty getting complainants to agree to mediation. Other oversight systems offer no mediation option.

The number of paid staff among the oversight systems examined ranges from 1 part-time person (St. Paul) to 30 full-time staff (San Francisco). Most systems have between four and five paid staff. Largely reflecting the number of paid staff, the systems' budgets range from almost no special funding in Orange County to $2,198,778 in San Francisco. While four other budgets range from $100,000 to $275,000, St. Paul's is slightly more than $37,000 and Minneapolis' is slightly more than $500,000.

Exhibit 2–2 illustrates the significant diversity in the type and extent of oversight activity levels. For example, San Francisco received 1,126 complaints in 1997, and 715 citizens contacted the Minneapolis program. Berkeley investigated 42 complaints. Orange County held 45 hearings in 1997, Berkeley 12, and Minneapolis 3. However, Minneapolis oversight staff also provided other assistance to 715 additional citizens.

Exhibit 2–2 shows that the systems' openness to the public also differs widely. At one extreme, Rochester reviews completed investigations in a private, sealed-off room in the city hall basement—even the citizen panelists have no access to internal affairs reports until the panelists meet, and they must surrender the materials before they disband. The panelists' findings are not made public. St. Paul's reviews also are conducted in private. By contrast, Orange County invites 57 media outlets to every board hearing, and the board members' findings are announced. Some systems mix privacy with openness. Flint's ombudsman's office conducts its investigations in private but then provides its detailed findings to the press and the city hall's public archives. The chief's hearings in San Francisco are private, but police commission trials have been attended by as many as 600 citizens and members of the press. Boards in Berkeley, Minneapolis, and Tucson conduct public meetings at which individual citizens can raise general concerns (not personal complaints) about police conduct. The boards then take up the concerns between meetings or at a future meeting. No oversight system publicizes the nature of the specific discipline subject officers receive—many jurisdictions prohibit such disclosure.

This tremendous variation in how the nine oversight systems conduct business may seem discouraging: The lack of similarity makes it difficult for other jurisdictions to make an automatic selection of commonly implemented oversight features around which they can structure their own oversight procedures. This diversity forces jurisdictions to take the time to pick and choose among a wide range of alternatives for designing their own oversight systems and to assess the benefits and limitations of each possible component. On the positive side, this diversity means jurisdictions do not have to feel they are obligated to follow rigorously any one model or approach; they have the freedom to tailor the various components of their system to the particular needs and characteristics of their populations, law enforcement agencies, statutes, union contracts, and pressure groups. Of course, the choices that are made may have important consequences for how much the oversight system will cost, how much it is utilized, and how satisfied citizens are with the complaint process—considerations that will in turn partially determine which options to select.

Although the choices may be daunting, there is expert help available for making them. Key participants in all nine oversight systems have agreed to field telephone calls from interested parties to share information about what works best for them and why. The names and telephone numbers of these individuals follow each case study. Chapter 8, "Additional Sources of Help," identifies other individuals with national experience with oversight systems who are available for consultation.

The Berkeley, California, Police Review Commission: A Citizen Board and the Police Department Investigate Complaints Simultaneously

Background

After allegations of police use of excessive force in clearing street people from a local park, Berkeley voters in 1973 approved a ballot initiative that created by ordinance the Police Review Commission (PRC), the oldest continuously operating citizen oversight agency in the Nation.

Citizens filed 42 cases with PRC in 1997. The board conducted hearings in 12 cases, which sometimes included multiple allegations (some of which came from the previous year's filings). The board sustained at least 1 allegation in 2 of the 12 hearings, for a total of 4 sustained allegations. The board did not sustain 30 allegations. The board closed another 34 cases without hearings, either because the case lacked merit or the complainant failed to cooperate. For the first half of 1998, in 5 of the 11 hearings held, there was at least 1 sustained allegation.

The review process

Exhibit 2–3 illustrates the Police Review Commission's procedures.

THUMBNAIL SKETCH: BERKELEY

Model: citizens investigate (type I)

Jurisdiction: Berkeley, California

Population: 107,800

Government: city council/city manager

Appointment of chief: city manager nominates, city council approves; city manager can remove

Sworn officers: 190

Oversight funding: $277,255

Oversight staff: two full-time professionals; two full-time clerical

Oversight supervisor: city manager appoints Police Review Commission officer

A nine-member, all volunteer Police Review Commission (PRC) appointed by the city council holds public hearings of citizen complaints against the police department (with three commissioners participating in each hearing). A PRC officer appointed by the city manager forwards each complaint she receives to the police department's internal affairs (IA) bureau, and she and IA conduct simultaneous but independent investigations of the complaint. The PRC officer forwards her investigation results to the PRC board for a hearing. After the hearing, the board submits its findings to the city manager and the chief. Any citizen may express concerns about department policies or procedures at full commission meetings. Based on these public meetings and examination of complaints citizens have filed, PRC recommends policy and procedure changes to the city manager and chief.

CHAPTER 2: CASE STUDIES OF NINE OVERSIGHT PROCEDURES

Exhibit 2–3. Citizen Complaint Process in Berkeley

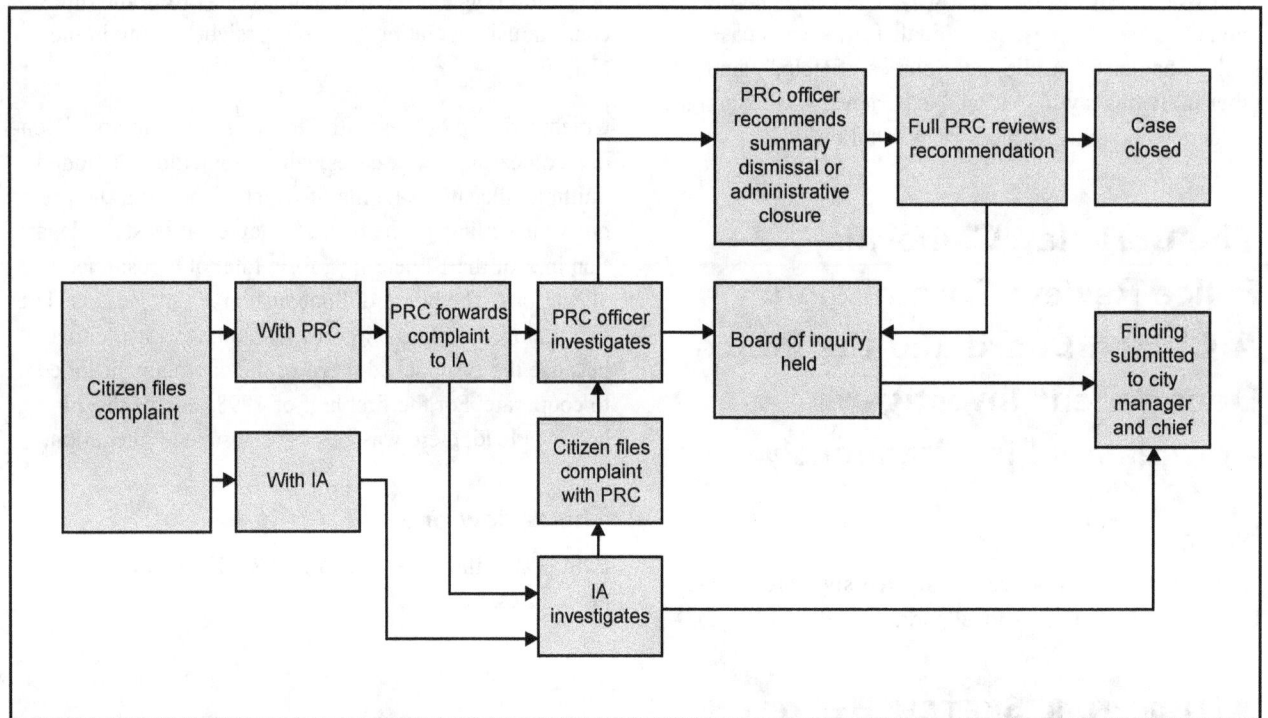

Intake
Citizens may file complaints directly with PRC within 90 days of the alleged misconduct. The ordinance requires PRC to forward complaints to the police department's internal affairs bureau within 30 calendar days. IA and PRC then both investigate the case independently. PRC and the police department have 120 days to communicate their findings to the city manager and for the city manager or chief to determine discipline.

Citizens who file a complaint initially with the police department's internal affairs bureau may file the complaint subsequently with PRC within the 90-day limit, after which the parallel PRC and IA investigations occur. The IA investigators give complainants a brochure on the complaint process that mentions PRC, and they tell citizens who express dissatisfaction with the IA investigation about the PRC option. From 1994 through 1998, 53 percent of complainants registered their complaints initially with IA rather than being referred by PRC.

Investigations
Either Barbara Attard, the PRC officer, or the PRC investigator conducts an investigation of each complaint. Subject officers must appear and answer questions, but they may appear with a union representative or lawyer.

Hearings
To hear each complaint, PRC staff impanel a board of inquiry consisting of three of the nine board members. The three choose a chairperson from among themselves. One week before the hearing, PRC staff provide the members with a packet containing the results of their investigation along with relevant ordinances, statutes, and department policies and procedures. Attard sends a notice to the chief who, according to the ordinance, must order the involved officer(s) to attend. A lieutenant, the duty command officer for the week, is always present to answer questions about police policy, procedures, and training.

As soon as the hearing begins, the chairperson makes clear that the board can offer only recommendations to the city manager and the chief. The hearing then proceeds as follows:

1. The complainant presents the complaint and introduces any witnesses.

2. Board members, and subject officers or their attorneys, may question the complainant and witnesses.

3. Steps 1 and 2 are followed for the subject officer.

4. Each party may make a closing statement.

5. The board deliberates in closed session.

6. The board returns to announce its finding.

According to the ordinance, the parties may present evidence "on which reasonable persons are accustomed to rely in the conduct of serious affairs," including hearsay. The chairperson rules on objections, but other board members can overrule the chair. (See "A Hearing by Berkeley's Police Review Commission.")

A Hearing by Berkeley's Police Review Commission

The chairperson calls the meeting to order at 6:10 p.m. A complainant has alleged that (1) an officer unlawfully taped a telephone conversation with her and (2) failed to give proper explanation by stating that her complaint was a civil, not a criminal, matter. The police officer's attorney begins by asking the panel to dismiss the case on procedural grounds because the officer was an IA investigator at the time of the incident. The chairperson refuses.

The chairperson then invites the complainant to state her complaint briefly. The woman describes her call to the police after a business partner jumped her locked fence, banged on her door, and demanded payment for an overdue bill. The complainant asks to play the tape recording of her original 911 call, but the chair rules the tape is irrelevant to this officer's case. (The complainant has filed a complaint against another officer in which the tape is pertinent.) The chair refuses two more requests by the officer's attorney to dismiss the case. On one of the attorney's motions, he says he will get an opinion from the city attorney. The officer's attorney asks the complainant several questions, after which two board members ask her questions.

The officer (who comes in uniform) is sworn in, but he says he has no statement to make. The complainant asks the officer several questions, including, "Don't you feel bad about not protecting me [by coming out to her home when she reported the trespass that was the origin of the case]?" The officer's lawyer objects to the question, and the chair tells the complainant to save these kinds of statements for her closing argument.

A commissioner asks the officer, "Do you tell people you are taping them?" "Usually," he responds. The chair then asks the lieutenant, "Does the department have a policy to record conversations and tell people whether they are taping them?" The lieutenant says there is no rule, but the practice is usually to tape and tell. Complainant: "May I say something?" Chair: "No."

The chair asks the officer why he did not tell the complainant to file a criminal complaint and let the district attorney decide whether the incident was a criminal matter. The officer shrugs his shoulders. The lawyer then asks the officer questions and gives her closing statement, repeating her three motions for dismissal. The complainant, too, gives a concluding statement, saying, "I am at a disadvantage here because the officer has an attorney, but I cannot afford one." She adds that it is insulting for the officer and his attorney to be chewing gum throughout the entire proceeding. The officer gets up and throws out his gum; his attorney does not.

The board leaves to deliberate at 7:15 p.m. and returns at 7:43 p.m. The chair reports that the board voted 3 to 0 not to sustain the first allegation (illegal tape recording) and 2 to 1 not to sustain the second allegation (improper advice).

As the meeting breaks up, the complainant tells the chair that she is very upset; board members remain about 5 minutes longer to listen to her frustrations with the hearing process and outcome. The PRC officer explains to the complainant her right to appeal the decision.

Findings

Board of inquiry findings are based on clear and convincing evidence. Possible findings include unfounded, exonerated, not sustained, and sustained. PRC presents its findings to the city manager and the chief. If the IA and PRC findings differ, a designee of the city manager reviews the decision and recommends to the city manager which finding to support. However, because IA has already completed its investigation and recommended a finding to the chief, the chief has typically already ruled on IA's finding and, if appropriate, imposed discipline. Nevertheless, because the city manager has ultimate authority in disciplinary matters, he can overturn the chief's decision after reviewing PRC's finding. In practice, however, the city manager does not try to reconcile different findings; the chief alone decides whether to reverse IA's finding.

Appeals

Within 15 days after the complainant and subject officer have been sent PRC's finding, either party may petition in writing for a rehearing. There had been no rehearings as of October 1998 because police officers had never requested one and complainants had not been able to document that they had newly discovered evidence.

Commissioners and the PRC officer have no regular procedure for learning what IA's dispositions are. California statute (§832.7) provides that "Peace officer personnel records and records maintained by any state or local agency, . . . or information obtained from those records, are confidential and shall not be disclosed in any criminal or civil proceeding except by discovery." As a result, the complainant does not learn whether or what kind of discipline, if any, the chief imposes. (See "A Citizen Has Mixed Feelings About the PRC Process.")

Other activities

PRC performs two additional functions.

Public forum for complaints and policy issues

At its general meetings held on the second and fourth

A CITIZEN HAS MIXED FEELINGS ABOUT THE PRC PROCESS

A Berkeley resident was stopped by a police officer and cited for a traffic violation. The citizen felt he had not committed the violation but had been singled out because of his ethnicity. A few weeks later, he heard about PRC from a friend, who had read about it in the newspaper. Several weeks later, he wrote a letter to PRC and the chief describing the incident and alleging several acts of misconduct by the officer.

An internal affairs investigator telephoned the man to say that IA would investigate the case independently of PRC. The IA investigator interviewed him on the phone. The PRC investigator interviewed the man face-to-face in a 1-hour taped interview. The PRC investigator warned him that the outcome of the case was uncertain. The complainant was frightened and tired and concerned the police might retaliate against him for having filed the complaint. But he followed through. After the interview, the investigator sent him a copy of the transcript along with the officer's statement.

At the hearing, the complainant and the officer each gave a statement and asked each other questions. The three board members asked them questions, too. The commissioners then left the room for 20 to 30 minutes to deliberate. Two board members found that the officer had engaged in an unprofessional backtalk, but all three exonerated the officer on the other allegations. The board member who dissented from the one negative finding explained his position.

The PRC investigator told the complainant at the end of the hearing that he would not learn whether the officer would receive any discipline for the sustained allegation. The complainant felt frustrated by this, but he also wanted to put the episode behind him. Overall, the complainant said, "If a similar incident happened again, I would still file a complaint with the PRC just to see justice done."

Wednesday of every month except August, PRC serves as a public forum at which citizens can express concerns about police policies and procedures. The meetings, announced in advance to the press, usually last about 90 minutes and draw as many as 30 residents and media if there is a controversial issue of community concern. An IA investigating sergeant attends every PRC general meeting.

At the meetings, the PRC chairperson asks for public comment, subcommittees (e.g., on community outreach) give reports, and new business is taken up. Barbara Attard gives a report on the number of new cases filed since the previous meeting and identifies cases that she recommends be closed administratively. She may invite a police unit (e.g., domestic violence, bicycle) to come to describe its activities.

PRC's charter also requires it to hold special public hearings at the request of board members or voters to air controversial matters related to allegations of police misconduct.

Policy recommendations
Either as a result of a public meeting or because of specific citizen complaints that PRC has heard, board members and Attard recommend changes in department policies and procedures. In the wake of riots in a local park in 1991, which resulted in over 30 complaints to PRC alleging officer misconduct, the city council directed PRC to review and make recommendations on "all aspects of crowd control policies at large demonstrations." After study and deliberation, PRC recommended 12 specific changes that the department later implemented (see chapter 3, "Other Oversight Responsibilities").

Staffing and budget

Each of the nine city council members appoints one PRC board member. Board members may serve indefinitely until the appointing city council members replace them. Most serve 5 or 6 years; four have served for at least 10 years. Board members select one of their members to a 1-year nonrenewable term as chairperson.

The city manager appoints the PRC officer and provides an investigator. Officially, the PRC officer and investigator are the city manager's staff. However, the public sees them as acting as the PRC commissioners' staff. The city manager in effect delegates his role in supervising PRC to the PRC officer. Two office assistants complete the staff.

PRC's budget for fiscal year 1998–99 was $277,255 (see exhibit 2–4). Until 1998, the budget declined steadily for several years, along with a reduction in staffing levels from six full-time equivalents (PRC officer, two investigators, three clerical support) in 1992 and 1993 to four full-time staff (two professionals and two clerical support) in 1997.

Distinctive features

Berkeley's oversight procedure is unusual in that the oversight body and the police department investigate many complaints simultaneously and independently, rather than sequentially. The system has other interesting features.

- Because the police department's IA unit and PRC conduct parallel investigations, if a citizen files a complaint with PRC, the case has the benefit—but incurs the expense—of two separate investigations.

- Although PRC must refer all complaints that citizens file with the board to the police department for simultaneous investigation, internal affairs does not refer cases routinely to PRC. While the PRC ordinance requires IA to refer all its complaints, State law makes citizen complaints filed with IA confidential. As a

EXHIBIT 2–4. BERKELEY POLICE REVIEW COMMISSION BUDGET, FISCAL YEAR 1998–99

Budget Item	Funding Level
Employees	$180,713
Employee education program	586
Fringe benefits	70,659
Stipend–police commission	12,390
Office equipment/furniture	2,895
Facilities maintenance	410
Building and structure	1,120
Telephone	1,761
Pagers	103
Central duplicating	680
Supplies/accessories	2,408
Postage	1,045
Workers compensation	2,485
Total	**$277,255**

result, the PRC officer was trying to develop a referral process that would comply with the statute and the PRC ordinance.

- The chief normally reviews IA findings on cases and, as appropriate, hands out discipline before he or the city manager receives PRC's findings.

- PRC's twice monthly public meetings make it possible for any citizen to express concerns about police misconduct or policies and procedures. The hearings have resulted in PRC making significant recommendations to the department for changes in policies and procedures.

For further information, contact:

Barbara Attard
Police Review Commission Officer
Police Review Commission
2121 McKinley Avenue
Berkeley, CA 94704
510–644–6716

Dash Butler
Chief of Police
Berkeley Police Department
2171 McKinley Avenue
Berkeley, CA 94703
510–644–6568

The Flint, Michigan, Ombudsman's Office: An Ombudsman Investigates Selected Citizen Complaints Against All City Departments and Agencies

Background

Sweden first incorporated the ombudsman concept in its constitution in 1909 as a means of curbing governmental abuses and protecting citizen rights. Today, an ombudsman typically investigates unlawful or unfair acts on the part of government agencies and complaints about their services.

THUMBNAIL SKETCH: FLINT

Model: citizens investigate (type 1)

Jurisdiction: Flint, Michigan

Population: 134,881

Government: strong mayor; city council

Appointment of chief: mayor appoints

Sworn officers: 333

Oversight funding: $540,744 (includes overseeing complaints against all city agencies)

Oversight staff: seven full-time professionals (two exclusively handle complaints against the police), one full-time secretary

Oversight supervisor: city council

The Flint, Michigan, City Office of the Ombudsman investigates complaints from residents dissatisfied with any city agency, but about half of its complaints are filed by citizens concerned with police officer behavior. The office settles some complaints by providing citizens with information about police department policies and procedures or through informal mediation. In serious cases, office investigators interview complainants and witnesses and require written answers to questions by subject officers. The office submits a report on each investigated complaint to the chief, who arranges for an internal investigation before deciding on a finding. The ombudsman's principal power lies in its ability to criticize openly the behavior of officers by name to the press.

In 1974, Flint voters adopted a new charter establishing an Office of the Ombudsman along with a strong mayoral form of government. Because some citizens felt a stronger mayor would need some checks and balances, the electorate simultaneously voted to include the ombudsman's office in the new charter for a 5-year period. In a 1980 referendum, nearly 60 percent of the residents voted to continue the ombudsman's office indefinitely.

The Flint City Charter states that "The Ombudsman may investigate official acts of any agency which aggrieve any person." City departments are required to provide information the ombudsman requests, and the office has the power to subpoena witnesses (including police officers), administer oaths, and take testimony. If elected officials or appointees refuse to cooperate, the charter provides for an obstruction hearing that could result in their forfeiting their jobs.

The ombudsman establishes his or her own rules for receiving and processing complaints, conducting investigations and hearings, and reporting findings. In 1996, the ombudsman's office investigated 662 cases, 313 of which (47 percent) involved complaints against police officers. In 1995, 389 of 741 cases (52 percent) involved complaints against the police. The office sustains 2 to 4 percent of citizen complaints against the police annually.

The review process

Exhibit 2–5 shows the process the ombudsman's office uses to review complaints.

Intake
People learn about the ombudsman's office from high-profile cases covered by the media or by word of mouth from coworkers. The police department's IA unit does not inform citizens about the ombudsman unless they report they are unsatisfied with the department's answers to their questions. In addition, the ombudsman's preferred response to complaints is to refer them to the appropriate supervisor, accepting complaints primarily when the citizen does not want to file with the police department or is dissatisfied with the supervisor's response, or when the complaint appears to involve the use of excessive force. Citizens who want to file complaints with the ombudsman must agree to be interviewed at the ombudsman's office or at a location of their choosing. The ombudsman assigns the citizens to one of two investigators who specialize in police complaints.

Informal resolutions
The assigned investigator may telephone the IA commander to resolve the complaint informally, such as clarifying a policy or procedure and then providing the explanation to the complainant. The IA commander may also choose to ask the shift commander of the subject officer to investigate the problem and then explain the officer's behavior to the complainant. About one-quarter of complaints reported to the ombudsman are settled by means of these informal approaches.

The ombudsman office investigator's next option is mediation. If both parties agree, the investigator arranges a

Exhibit 2–5. Flint Office of the Ombudsman's Investigation Process

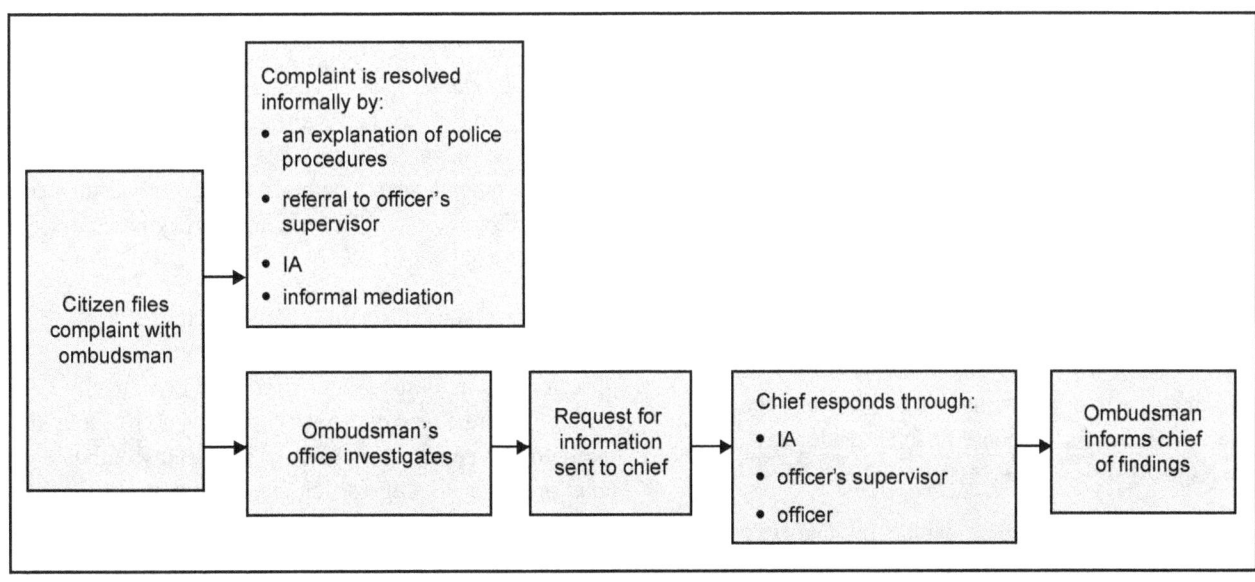

meeting through the officer's supervisor, if necessary walking the complainant to the police department to talk with the supervisor. The citizen and supervisor meet together alone. If the citizen is not satisfied, he or she then can file a complaint with the ombudsman.

Formal investigations
When the ombudsman's office accepts a complaint, the investigator sends the chief a letter reporting the complaint and asking for a response to questions from the officer. The chief sends the letter down the chain of command to the subject officer, who usually responds to the questions in writing or, on rare occasions, in an interview.

The investigator also interviews the complainant for his or her account of the incident and the names of witnesses. Investigators usually tape the interview. The investigator attempts to contact witnesses by telephone and, where appropriate, sends letters to homes in the immediate area of the incident. As needed, the investigator also takes photos at the scene, secures medical records, and undertakes other pertinent investigatory activities. The ombudsman's office has never subpoenaed a witness.

The investigator turns in a report to the chief investigator or deputy ombudsman indicating agreement or disagreement with the citizen's allegation(s). The investigator meets with the deputy or ombudsman to decide on a finding.

Findings
The ombudsman's office either sustains or does not sustain each allegation, sustaining only if there is clear and convincing evidence. The office sends a complete report of each investigation to the chief and the city council. The office recommends whether there should be discipline but not the type of discipline.

When the ombudsman's office concludes the officer did something wrong—which happens 5 to 10 times a year—it sends the officer and the chief a synopsis of its investigation with its conclusion. The chief then conducts his own investigation through IA or the officer's commander and makes a final determination of how to proceed. (See "The Chief's Response to an Ombudsman Investigation.")

The chief sends the ombudsman his finding. He does not inform the ombudsman's office about IA's finding, and he has the discretion not to tell the office whether he imposed any discipline. However, on occasion the city council has asked the chief to explain his response to an ombudsman's report.

The ombudsman's investigator telephones or writes each complainant to report the chief's decision. The typical case is resolved in 3 weeks.

Other activities

Because there is no shield of confidentiality in Michigan, the ombudsman's office has considerable latitude in informing the press about its cases and criticizing officers by name. The office routinely sends its case reports to the city clerk as public documents for the city archives. However, the city charter requires that "No report or recommendation that criticizes an official act shall be announced until every agency or person affected is allowed reasonable opportunity to be heard with the aid of counsel." As a result, the ombudsman's office circulates the report on every sustained complaint to everyone named in the report (except the complainant), giving them 5 days in which to challenge its factual accuracy (but not the findings).

Staffing and budget

By a two-thirds majority of the nine members, the city council appoints the ombudsman for a single 7-year term. A three-quarters majority on the council can remove the ombudsman.

At one time, the office had as many as nine investigators, but by 1998 the number had declined to five. Two investigators handle police complaints full time, and the deputy investigator takes on some police complaints as well. The ombudsman appoints a deputy ombudsman and the investigators. The office has an attorney on contract to answer legal questions.

There was no ombudsman's office director between August 1995 and the end of 1998. When the previous director was fired in 1995, a court ruled that the city could not hire a new director as long as a civil suit by the fired employee was still pending. The deputy ombudsman or senior investigator ran the office in the absence of a director. In September 1998, a Michigan appeals court ruled that the city could hire a new director.

THE CHIEF'S RESPONSE TO AN OMBUDSMAN INVESTIGATION

A man arrested on a domestic violence charge filed a complaint with the ombudsman's office alleging an officer punched him in the face after the citizen tried to headbutt the officer. The citizen also alleged that the officer threatened to beat him up for hitting a woman.

An ombudsman's investigator interviewed the complainant, his girlfriend, and the complainant's brother as well as the subject officer and two other officers at the scene. Medical records indicated a 2 by 2-centimeter hematoma on the right cheek. The citizen's girlfriend reported, "I heard him [the officer] say [to the citizen on the phone that] he was gonna kick his a--." The citizen's brother reported that "I heard one cop say that he [the complainant] hit [the officer]."

The subject officer wrote to the ombudsman that some words were exchanged between him and the citizen, and when the citizen said "F--- you" and headbutted him, the officer immediately struck the man with a closed fist to the face. The other officers reported that the man was already handcuffed at the time the officer hit him. The officer said he hit the man because "With the quickness of the situation, I had no time to use my O.C. [oleoresin capsicum, or pepper spray] or any other methods to control [the man] from striking again."

The ombudsman's report sent to the chief reproduced two department policies pertinent to the complaint, one on self-control and one on the use-of-force continuum. The latter policy includes the statement that "Above the holds and maneuvers [in the continuum of force] are the STRIKINGS. The striking points may be soft tissue, joints, or, in the extreme case, the suspect's head."

The ombudsman's office summarized the case by saying:

> Other methods available [to the officer for restraining the subject] would include verbal persuasion, touching or pushing away, O.C. spray, a compliance hold, the assistance of the other two officers to subdue Mr. [the complainant] ... or simply stepping away from Mr. ____ to deescalate the altercation. It is the Ombudsman's determination that Officer ____ could have used any of the above mentioned alternatives first, rather than punching Mr. ____ in the face. Officer did not indicate in any reports to the Ombudsman that he felt his safety or life were in danger.

The report concluded by saying that "Officer ____'s actions violated the Flint Police Department' Use of Force Continuum ... [and] the police department's policy on self control.... Chief of Police Trevor Hampton should review Officer ____'s actions and issue the appropriate discipline."

The chief wrote the ombudsman's office back as follows:

> I am in receipt of your critical report. Please be advised that Mr. ____ did not file a complaint with the Flint Police Department regarding this incident. As a result of your report, I am initiating an investigation. If the findings of the internal investigation show violations by members of the Flint Police Department, appropriate action will be taken.

Three months later, the chief wrote again to say:

> "The investigation involving the complaint of Mr. ____ has been reviewed and evaluated by me. The charge has been sustained against Officer ____ and appropriate disciplinary action will be taken."

As shown in exhibit 2–6, the ombudsman's 1998–99 budget was $540,744; 91 percent of the budget represented wages and benefits. With two investigators devoting nearly full time to complaints against the police, and the deputy devoting about one-quarter time to police cases, the proportion of the budget devoted to complaints against the police is about $174,000.

Distinctive features

Few jurisdictions in the country make use of an ombudsman to review police misconduct complaints.

- Because the ombudsman serves as a generalized complaint handler for all government agencies, the city cannot be criticized for singling out the police for oversight.

- The ombudsman's office provides citizens with an alternative place to file complaints against the police department.

- The ombudsman's office helps IA to address complainants' concerns by offering a satisfactory explanation for an officer's behavior that the complainant could not or would not get from the subject officer or patrol desk.

- The office can subpoena department heads, including the chief, as well as employees and all case files. It has never used this power.

- The office can—and does—criticize officers by name in the media for their behavior. This may serve to deter some misconduct and anger officers. The public has the opportunity to become aware of police misconduct when the press prints the information.

- Politics could emasculate the office. Because the mayor appoints the chief and the city council appoints the ombudsman, conflict between the two could stymie the office's leverage if the mayor were to choose to ignore the ombudsman whenever the ombudsman wished to take serious exception to a chief's findings.

For further information, contact:

Jessie Binian
Ombudsman
Office of the Ombudsman
City of Flint
Flint Municipal Center North Building
120 East Fifth Street, Second Floor
Flint, MI 48502
810–766–7335

EXHIBIT 2–6. FLINT OMBUDSMAN'S OFFICE 1998–99 BUDGET

Budget Item	Funding Level
Wages and salaries	$273,639
Fringe benefits	220,123
Supplies	5,784
Newspapers, professional dues, and publications	240
Professional services	20,000
Micro software and leases	1,000
Data processing services	3,358
Professional services and commissions	3,400
Communications	2,500
Transportation	900
Printing and publishing	1,500
Insurance and bonds	100
Repairs and maintenance	2,000
Miscellaneous	200
Education, training, and conferences	6,000
Total	**$540,744**

The Minneapolis Civilian Police Review Authority: An Oversight System Investigates and Hears Citizen Complaints

Background

The Minneapolis city council established the Civilian Police Review Authority (CRA) by ordinance in 1990 after African-American community leaders led protests at city hall because officers had killed an elderly African-American couple in a raid and had broken up an apparently peaceful African-American college student party in

THUMBNAIL SKETCH: MINNEAPOLIS

Model: citizens investigate (type 1)

Jurisdiction: Minneapolis, Minnesota

Population: 358,785

Government: strong mayor, city council

Appointment of chief: mayor nominates, city council approves

Sworn officers: 919

Oversight funding: $504,213

Oversight staff: seven full time

Minneapolis' Civilian Police Review Authority (CRA) operates in two stages:

1. Paid, professional investigators and an executive director investigate citizen complaints to determine whether there is probable cause to believe misconduct occurred.

2. Volunteer board members conduct hearings to determine whether to sustain the allegations in probable cause cases.

In 1998, subject officers stipulated to a sustained finding in about half of the cases in which the CRA executive director found probable cause. CRA arranged for successful mediation in another 14 cases. As a result, only 10 hearings were held in 1998.

---

a Minneapolis hotel. In 1997, the city council and the mayor saw the need to determine whether CRA was providing the appropriate oversight in the most cost-effective manner and if it had the structure and staff to do so. As a result, they appointed a redesign committee that held focus groups, took public testimony, looked at how other jurisdictions configured their citizen oversight procedures, and then recommended changes in how CRA operated, most of which the city council and the mayor adopted.

In 1997, 715 individuals contacted CRA with concerns about possible police misconduct. Of these, 114 were satisfied with an explanation of the police department's policies and procedures. Another 87 callers were satisfied when investigators called the subject officers' supervisors to resolve the complaint. In 332 cases, there was no basis for a complaint, the caller was referred elsewhere, or citizens failed to follow up their initial reports. Twenty-three cases were pending.

The remaining 159 individuals signed formal complaints (see exhibit 2–7). Of these, the CRA executive director found no probable cause in 46 cases because of insufficient evidence. The executive director exonerated officers in another 54 cases because the facts in the allegations were untrue or, while true, did not constitute misconduct. The executive director dismissed another 30 cases, for example, because the complainant failed to cooperate. Fourteen cases were successfully mediated, and five cases were pending as of the end of the year. Of the 10 cases in which the executive director found probable cause, 9 were sustained, 6 by stipulation (see next section) and 3 at hearings. One case was still pending at the end of the year.

CHAPTER 2: CASE STUDIES OF NINE OVERSIGHT PROCEDURES

EXHIBIT 2–7. DISPOSITION OF 159 SIGNED COMPLAINTS IN 1997

Insufficient evidence	46
Exonerated	54
Dismissed	30
Mediated	14
Pending	5
Probable cause	10
Sustained by stipulation	6
Sustained at hearings	3
Pending	1
Total	**159**

The complaint process

Exhibit 2–8 is a flow chart that summarizes how CRA processes cases. The following discussion explains each step.

Intake

When complainants contact CRA or the police department's IA unit, they are told they have the choice of filing with either office but not both. Furthermore, if they are unsatisfied with the finding from one office, they may not then file with the other office. Only IA handles allegations of misconduct that require a criminal investigation, could lead to an officer's being fired, or are high profile.

If the complainant files with CRA, the secretary assigns an investigator who sees the walk-in immediately or telephones the caller to set up an appointment to meet at CRA. The investigator fits each charge the complainant alleges into one of eight general CRA types of complaints

EXHIBIT 2–8. MINNEAPOLIS CIVILIAN POLICE REVIEW AUTHORITY COMPLAINT PROCESS

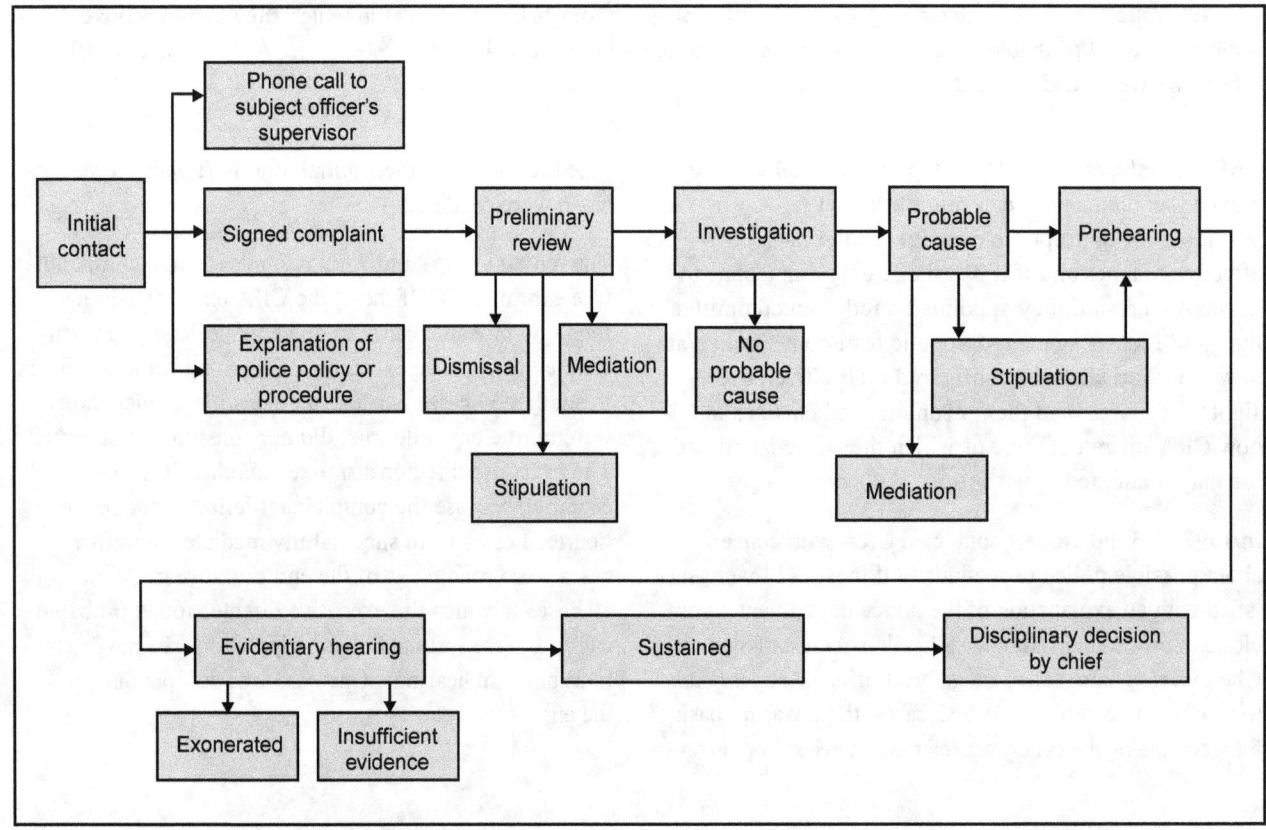

(e.g., excessive force, inappropriate language). The investigator then identifies the department's policy or procedure that appears to have been violated. The executive director may dismiss the case during this preliminary investigation stage.

The investigator sends a letter to the complainant with a copy of the complaint form asking the person to correct any errors and sign and return it within 15 days. The executive director sends a notice of the complaint filing to the officer, deputy chief, and chief. When the investigator wants to take a formal statement from the officer, the chief sends a *Garrity* warning (see "Glossary") requiring the officer "upon pain of disciplinary action" to make an appointment with CRA to answer questions.

CRA investigations

The investigator interviews any witnesses the complainant may have identified and does any additional needed leg work, such as confirming visually that a witness had an unobstructed view of an incident from her bedroom window and enough street lighting to see the nighttime activity clearly. Investigators have gone door to door in neighborhoods leaving business cards for potential witnesses.

Investigators interview the subject officer last. About half the officers bring a union representative or attorney, who may caucus with the officer but not speak. The investigator sends two copies of a transcript of the taped interview to the officer, one of which the officer signs and returns. At the conclusion of the investigation, the investigator forwards the file to the executive director that includes the investigator's conclusion regarding the probable cause for each allegation. If the recommendation is that there is no probable cause, the investigator recommends a finding of either insufficient evidence or exoneration. If there is probable cause, the investigator cites the policy or procedure that the subject officer appears to have violated. Patricia Hughes, the executive director, makes a final determination regarding probable cause.

At this stage, but sometimes before the probable cause finding or during or after the prehearing (see next section), the officer and CRA executive director may strike the equivalent of a plea bargain, with the officer stipulating to one or more allegations (that is, admitting guilt) in exchange for CRA dropping one or more other allegations (see "Stipulations Reduce CRA's Caseload"). The complainant is not consulted regarding the nature of the stipulation.

If there is no stipulation and no offer and agreement to mediate (see "Other CRA activities" on page 34), the CRA chairperson appoints a three-member panel to hear the case, designating one of the members (including himself, if he so chooses) as the panel chair.

STIPULATIONS REDUCE CRA'S CASELOAD

Patricia Hughes, the CRA executive director, initiated stipulations after a police union representative suggested that his client would agree to having committed one allegation if CRA would drop the other complaints. Because the number of hearings had created an accumulation of pending cases, Hughes saw stipulations as an opportunity to reduce the backlog.

After a stipulation, CRA informs the chief of which allegations were sustained and tells IA that the findings are the result of a stipulation. Officers have never agreed to stipulate in cases of alleged use of excessive force.

Officers and their representatives have learned that if the executive director decides there is probable cause that the officer committed the alleged misconduct, a CRA panel will sustain all of the allegations in about three-quarters of the cases. As a result, it is usually in the officers' best interest to agree to a stipulation to get some of the allegations dropped.

Because officers have been increasingly willing either to stipulate or agree to mediation, there were only five hearings from January 1, 1998, through November 30, 1998.

The evidentiary hearing

The subject officer appears with a union attorney at the CRA office for a half-hour "prehearing" at which the panel chairperson and the CRA executive director agree on the witnesses they will be bringing to the evidentiary hearing and the information and materials that each side will be permitted to introduce, such as the incident report, medical records, and department training manual. The panel chairperson rules on what information may be introduced. The prehearing makes it possible to avoid spending time at the hearing deciding what type of evidence will and will not be admissible.

Based on the union attorney's and executive director's schedules, the panel chairperson schedules a hearing 6 to 8 weeks after the prehearing. Panel members (except the chairperson, who has attended the prehearing) know nothing about the case until the hearing begins.

Each panel holds a private, audiotaped evidentiary (administrative) hearing lasting from a few hours to, on occasion, several days. Patricia Hughes, CRA executive director and a former assistant city attorney, prosecutes the case, and the police union lawyer defends the officer. CRA does not have subpoena power, but officers must testify under the *Garrity* ruling. After witnesses are sworn in, each side questions its witnesses, who are then cross-examined by the opposing side (followed by recross). The prosecutor explains why she believes the officer's behavior violates a department policy or procedure.

Panel members may question witnesses and usually do. The chairperson rules on any objections raised by the union attorney or executive director. The panel may admit all evidence that furnishes proof of guilt or innocence, including reliable hearsay if it is the type of evidence that "reasonable persons are accustomed to rely on in the conduct of their serious affairs." While the officer remains during the entire hearing, the complainant leaves the hearing after giving his or her testimony because the Minnesota Data Practices Act give employees (e.g., officers) privacy in administrative hearings. The prosecutor may present a final rebuttal to the union attorney's closing statement.

Findings

The panel deliberates in private, using a clear and convincing standard to sustain or not sustain the complaint(s). The CRA executive director sends a letter to the subject officer and deputy chief presenting the panel's finding. Within 5 days of receiving the panel's finding, the officer or complainant may write to ask the panel to reconsider its finding. About 5 percent of cases are appealed; few appeals are granted.

The police department's disciplinary panel reviews CRA's finding and recommends discipline. The officer may appear before the disciplinary panel with a union representative to challenge the offense severity but not to contest the CRA finding. The panel forwards its disciplinary recommendation to the chief for final review. The ordinance requires that the chief decide on discipline based on the results of the hearing and then, within 30 days, provide CRA and the mayor with a written explanation of the reasons for his or her action. The chief may not reverse a CRA finding but has the authority to decide whether to punish the officer and what discipline to impose. (See "A Sample Hearing.")

Other CRA activities

CRA has three other responsibilities.

Monthly CRA meetings

The CRA board members and staff hold an open meeting the first Wednesday of every month at 5:00 p.m. in an office building. The executive director keeps the public apprised of CRA's activities, providing updates on the number of cases opened and resolved. The board asks if anyone in the audience wishes to express general concerns about police behavior. Patricia Hughes relates the following story:

> A few citizens expressed objections at one meeting to the manner in which officers were conducting apartment searches to find suspected drug dealers. In these instances—a tiny minority of all drug searches—the officers had raided the wrong address or the drug dealing had apparently been occurring while the legal tenants were not present. However, because the raids involved no-knock entries with shotguns and orders for everyone in the apartment to lie down at once on the floor (including a woman sleeping in the nude), the tenants had been embarrassed, frightened, and angry. I met with department inspectors to share the

A Sample Hearing

The chairperson called the hearing to order at 4:05 p.m. Patricia Hughes, the CRA executive director, acting as the prosecutor, began by giving an opening statement in which she described how an off-duty white officer, stationed at an upscale hotel, was alleged to have made a racial slur against a black man, not a hotel resident, for being messy while using the hotel restroom. The officer and the man got into a heated discussion, after which the officer "trespassed" the man—giving him notice that he would be subject to arrest if he returned to the property.

The citizen filed a complaint for harassment and inappropriate behavior. The union attorney said there had been numerous cases of vandalism and drug use in the hotel restroom. As a result, the officer was just doing his job to protect the premises in questioning the man about his behavior. The complainant was sworn in and answered questions from the prosecutor, union lawyer, and panel members. He then left.

The prosecutor introduced a friend of the complainant's who had entered the hotel with him but had not used the restroom. However, the friend had heard the conversation that transpired in the hotel lobby and confirmed the complainant's story. Although this was hearsay evidence, the panel accepted it because the witness was so close to the event in time and place. The officer was then sworn in and given a *Garrity* warning. The officer denied having made any racial slurs.

The hearing concluded with the prosecutor and union lawyer offering concluding statements, and the prosecutor presenting a final rebuttal to the lawyer's statement. The panel deliberated for about a half hour and found 2 to 1 for the complainant.

public's concerns. The inspectors agreed with my recommendation that the officers, when they fail to find the drug dealer, apologize to the tenants and explain that they had to take severe measures in order to protect themselves from drug dealers who are usually armed and often violent.

Mediation

After a citizen has filed a complaint against an officer, if both parties accept an offer to mediate, Patricia Hughes sends the case to the Minneapolis Mediation Program, a private, nonprofit organization with which CRA has a $1,500 annual contract to provide unlimited mediation services, typically 40 to 50 sessions a year. Under the terms of the contract, the program must arrange the mediation within 14 days unless there are extenuating circumstances. Mediation program staff telephone the parties to reconfirm they are willing to participate, explain the process, and set a time and neutral place (e.g., a library or neighborhood center) at the parties' convenience. The program informs Hughes whether or not mediation is successful. If mediation is successful, Hughes dismisses the complaint; if it is not, she sends the case back to her staff for investigation.

(See chapter 3, "Other Oversight Responsibilities," for additional information about the mediation process in Minneapolis.)

Early warning system

The IA unit generates a quarterly report that lists the 10 officers with the most complaints during the quarter and for the previous 12 months. The report distinguishes complaints filed with IA and those filed with CRA. CRA generates the totals for the complaints it receives.

Staffing and budget

By a majority vote, the city council appoints four board members through a public application process. The mayor nominates three board members as well as a chairperson from among the seven members. While the city council must approve the mayor's nominees, it has never rejected one. All appointments are for 4 years, subject to reappointment.

The CRA board hires, supervises, and fires (if necessary) the executive director. The CRA chairperson supervises and evaluates her. He asks other board members to fill out an evaluation on her each year and invites them to sit in on the in-person evaluation. He considers how she has managed the office as well as her litigation skills, public relations work, and timeliness (for example, whether she allowed too many extensions because she failed to supervise the investigators adequately). The executive director hires the three investigators—typically former police officers from other departments—and clerical staff.

The CRA's 1998 budget appropriation was $504,213 (see exhibit 2–9). More than three-quarters of the funding is for the salaries and benefits of seven staff: the executive director, three case investigators, a program assistant, and two clerk typists.

Distinctive features

The Minneapolis oversight system is unusual in that paid staff investigate most citizen complaints, while volunteers conduct hearings that result in findings the chief must accept. The system has several other notable features.

- Because the board appoints the executive director, she may be better shielded from political influence than if the mayor or city council appointed her. However, because the board hires the executive director, there could be a tendency on her part to accommodate the board's concerns rather than to act as a check and balance on each other (for example, when the executive director prosecutes cases before the board).

- Because most CRA investigators are former police officers, they have a good understanding of the nature of police work (see chapter 4, "Staffing"). At the same time, civilians with no professional experience as sworn officers conduct the hearings. As a result, CRA combines law enforcement and citizen perspectives.

- Using former police officers as investigators may result in bias in favor of officers; their use may also reduce the program's capacity for objectivity in the eyes of some citizens and community groups. Using former police officers as investigators may reduce opposition to the process among line officers and union leaders.

- Because the complainant may not attend the hearing except to give testimony and hear the attorneys' concluding statements, the complainant does not know why the case was won or lost.

- On one hand, offering stipulations reduces the number of cases CRA has to hear, which enables it to hear other cases more expeditiously. On the other hand, in some cases, stipulation can prevent mediation, when mediation might be useful as a procedure for educating the officer and the complainant to each other's points of view.

- By reducing the amount of time panelists have to spend at hearings deciding what types of evidence to allow, prehearings speed the process. Prehearings also offer another opportunity for subject officers to agree to stipulate as they reconsider the strength of the case against them.

- In its investigatory capacity, CRA is supposed to be a neutral party between the complainant and the police officer. However, if the case goes to a hearing, the CRA executive director prosecutes the officer. This dual role could confuse the public, complainants, and police officers.

For further information, contact:

Patricia Hughes, J.D.
Executive Director
Civilian Police Review Authority
City of Minneapolis
400 South Fourth Street, Suite 1004
Minneapolis, MN 55415–1424
612–370–3800

Exhibit 2–9. Minneapolis Civilian Police Review Authority 1998 Budget

Budget Item	Funding Level
Salaries and wages	$323,303
Benefits	68,518
Total personnel	**391,821**
Operating costs	33,169
Equipment	2,000
Contractual services	77,223
Total nonpersonnel	**112,392**
Total expenses	**$504,213**

Liz Murray
Mediator
Minneapolis Mediation Program
Hyatt Merchandise Mart
1300 Nicollet Mall, Suite 3046
Minneapolis, MN 55403
612–359–9883

Robert Olson
Chief
Minneapolis Police Department
Room 130, City Hall
350 South Fifth Street
Minneapolis, MN 55415–1389
612–373–2853

The Orange County, Florida, Citizen Review Board: A Sheriff's Department Provides Executive Support to an Independent Review Board

Background

In 1992, in response to the nationwide concern about police misconduct generated by the Rodney King beating, the Orange County Sheriff's Office established a process in which citizens could exercise oversight over deputies' use of excessive force and abuse of power. In 1995, the elected Orange County Commission amended the county charter to establish an independent Citizen Review Board (CRB) that effectively replaced the sheriff's board.

CRB heard 45 cases involving 67 allegations of misconduct that the sheriff's office investigated in 1997. The board disagreed with three IA findings, exonerating deputies of two allegations of abuse of power that IA had sustained and sustaining one abuse of power allegation in a case in which IA had exonerated the deputy.

The CRB procedure

Exhibit 2–10 and the following discussion explain the citizen oversight procedure in Orange County.

THUMBNAIL SKETCH: ORANGE COUNTY

Model: citizens review cases (type 2)

Jurisdiction: Orange County, Florida (Orlando)

Population: 749,631

Government: county commission

Appointment of sheriff: elected

Sworn deputies: 1,134

Oversight funding: $20,000

Oversight staff: two part time

A nine-person Citizen Review Board selected by the Orange County Commission and sheriff hears all cases involving the alleged use of excessive force and abuse of power after the sheriff's internal affairs unit has investigated them. Hearings are open to the public and the media. Board members also make policy recommendations. A captain in the sheriff's office devotes about 20 percent of his time to coordinating the board's activities.

Intake

Most citizens call the sheriff's office's internal affairs unit to file complaints, but others call the Citizen Review Board's number. The CRB's telephone number rings at the sheriff's Research and Development Unit, which has a dedicated line. The switchboard operator answers, "Citizen Review Board." When citizens call, the secretary mails out the CRB complaint form, which complainants return by mail or in person at the CRB office, located at the sheriff's office. The CRB secretary turns cases over to the sheriff's office IA unit for investigation and disposition.

Melvin Sears, a captain with the Research and Development Unit and the CRB administrative coordinator, provides board members with all completed investigations a month before the cases are to be heard. The cases are complaints of alleged use of excessive force (including all discharges of a firearm, even if there has been no citizen complaint) and abuse of power (using one's official position for personal gain or privilege or for avoiding the consequences of illegal acts). These types of cases are automatically slated for a future agenda. (See "A CRB Hearing Through the Eyes of a Deputy Sheriff.") Complaints that are questionable as to whether they fall within CRB's purview are given to the chairperson, and he decides if they are appropriate for board review.

CRB hearings

CRB meets once a month in public session in a county administration building meeting room. During the first part of the meeting, members approve the minutes of the previous meeting and hear any reports from the chairperson, vice chairperson, and Sears. The members then review cases in accordance with a published agenda that has been circulated in advance to the public and 57 media outlets. The board hears about four cases at each

EXHIBIT 2–10. THE ORANGE COUNTY CITIZEN REVIEW PROCESS

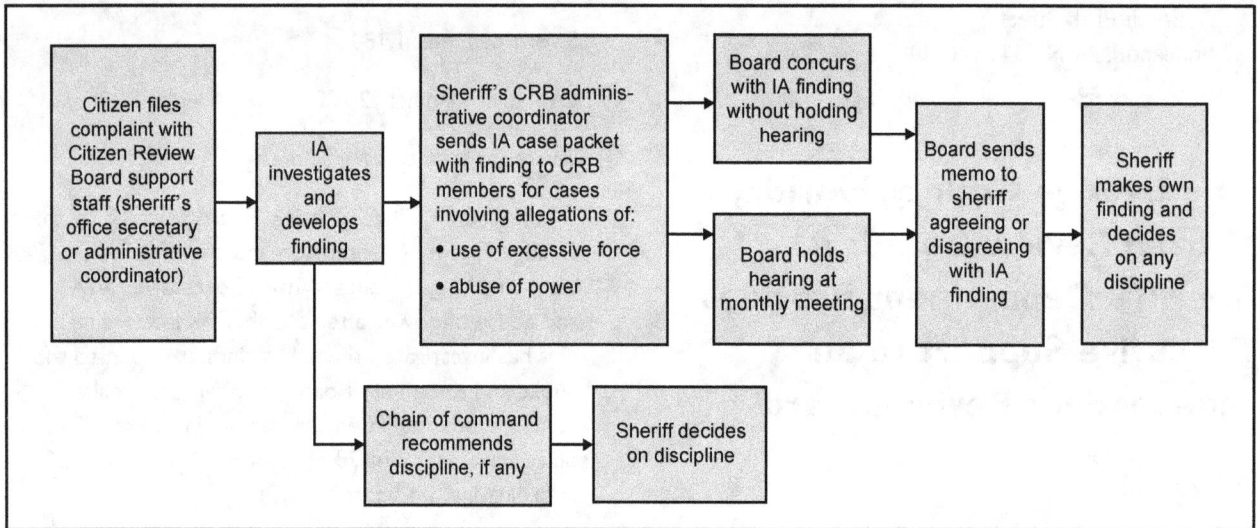

A CRB HEARING THROUGH THE EYES OF A DEPUTY SHERIFF

The IA unit told Patrick Reilly, a deputy sheriff, that the father of a youth Reilly had arrested had filed a complaint alleging use of excessive force in the form of a controlled knee spike (kick). Later, IA informed Reilly that it had exonerated him. However, the deputy knew that, because it was a use-of-force complaint, CRB would hold a hearing.

Within 2 weeks, CRB sent Reilly a letter instructing him to appear for a hearing and to bring any witnesses he chose. The deputy chose not to bring a union representative because there was not going to be a criminal charge and he felt confident he would be exonerated. Reilly did bring two other deputies who had witnessed the kick. Eight of the nine board members were present.

The chairperson called Reilly's case (four other cases were heard that evening), read the allegation, and asked for the deputy's side of the story. Reilly reports that he was given as much time as he needed and every opportunity to defend himself and clarify what he did and why. He did not feel he was on trial, and the board seemed neutral. The board asked one of Reilly's two witnesses to speak briefly. The IA investigator explained the sheriff's office use-of-force matrix and policy, which the board had already examined. The complainant did not come.

The board concurred with the IA finding. The case took slightly more than an hour. Reilly remained to sit in on the case that followed.

meeting. The IA investigating deputy is present at the hearing to answer questions about his or her investigation. Sears advises on policy issues and provides administrative support. The board has subpoena power, and county legislation provides for a fine of up to $500 or imprisonment for up to 60 days for anyone convicted of ignoring a CRB subpoena. However, the board has never subpoenaed anyone because the sheriff has issued a standing order requiring deputies to appear when called (but not requiring them to testify). The Central Florida Police Benevolent Association provides interested deputies with representation, but most deputies choose not to be represented because either they have decided not to answer questions or, more commonly, having already been cleared by IA, they feel they have nothing to fear. Finally, a criminal attorney, hired by the county on a retainer basis, comes to every hearing to answer questions on points of law, such as the proper interpretation of the State statute on assault.

Any board member may make a motion to place a complaint on a "consent agenda" if he or she feels that IA's findings are appropriate and no further review or meeting time is needed to discuss the merits of the complaint. Any member may also have a complaint removed from the consent agenda and subject to a full CRB review. The meeting minutes for the August 1998 meeting show, for example:

> Motion was made by Mr. Mills, seconded by Mr. Rankin and unanimously agreed upon, to place this case on the consent agenda, thereby concurring with the findings of the Professional Standards investigation that, based on a preponderance of the evidence, the following violations were sustained [the complaint and violation of policy followed].

Hearings follow Robert's Rules of Order. For each hearing, the following individuals, in this order, give a statement and answer questions from board members:

- The complainant.

- Any witnesses for the complainant (although they rarely appear).

- The sheriff's investigating agent.

- The subject deputy.

- Any other sheriff's employees present whom the chair chooses to recognize.

- The complainant again (to rebut the testimony presented by others).

At its discretion, the board may allow direct conversations among the parties.

Findings

The board spends 15 to 20 minutes deliberating each case in open session. The chairperson calls for a vote, and each member explains his or her decision. A majority rules, but almost all cases are unanimous. Decisions are based on a preponderance of the evidence. The board chairperson signs a form letter that Melvin Sears sends to each complainant after each hearing. There is no appeal.

CRB does not provide findings; rather, it sends a form memo to the sheriff agreeing or disagreeing with the IA finding in each case. The board agrees with IA findings 80–90 percent of the time. The board's decision is only advisory to the sheriff. On rare occasions, the sheriff overrules the board:

> A robbery detective on a stakeout fired a shotgun at a robber's car as it fled the scene, blowing out a tire. Because the sheriff's office prohibits firing at automobiles, IA determined that the deputy had violated department policy. The CRB, however, exonerated the deputy because members did not want to see deputies' hands tied so stringently—they wanted to provide deputies with more latitude in the use of firearms. Nevertheless, the sheriff supported the IA finding and disciplined the deputy.

Other activities

The board may recommend fitness-of-duty evaluations, additional training, and other measures for officers whose cases come before it. CRB also has the authority to hire an investigator to conduct its own investigations. However, when members feel more investigation is needed, they ask IA to do so and bring back the case. The board has recommended several policy and procedure changes that the sheriff has implemented (see the examples in chapter 3, "Other Oversight Responsibilities").

Staffing and budget

Each of the seven members of the Orange County Commission nominates a single board member subject to confirmation by the rest of the commissioners. The sheriff selects two members of the board—always choosing civilians. Board members serve for 2 years and may be reappointed for a total of 4 years. The members elect a chairperson for a year, who chairs every hearing.

The county commission requires the sheriff to assign a captain and a secretary unaffiliated with internal affairs to devote about 20 percent of their time to providing administrative support to CRB. Melvin Sears schedules CRB meetings and training sessions, informs complainants about the meetings, lines up the meeting hall, sends the IA case investigation materials to members, and prepares the annual CRB report. Sears keeps track of the board members' attendance on a spread sheet so he can report excessive absences to the county commissioners. The secretary records each hearing and provides the minutes.

Sears' and his secretary's CRB work amount to a $20,000 contribution by the sheriff's office (20 percent of their combined salaries). The sheriff also pays for all the direct costs associated with the board's work, such as postage and duplication. The sheriff pays for publishing CRB's brochure and letterhead stationary. CRB's attorney submits a bill to Sears—typically $200–$250 per month—who approves it and forwards it to the county for payment.

Distinctive features

An unusual feature of the Orange County oversight system is that a sheriff's deputy has the responsibility for administering the Citizen Review Board's activities. However, the board comes to its own conclusions in reviewing internal affairs findings. Because of this arrangement, the oversight procedure costs the taxpayer little.

- By keeping track of board members' attendance and reporting problems to the county commission, the administrative coordinator exercises some quality control over the proceedings.

- Locating the CRB office in the sheriff's office saves money otherwise needed to rent space and spares the administrative coordinator from having to shuttle back and forth between the agency and an outside CRB location. However, some complainants may be discouraged from filing because they are uncomfortable going to the sheriff's office.

- By reviewing all cases involving discharge of a firearm, regardless of whether a citizen filed a complaint, CRB can help identify problems among individual officers or general failures of training and policy.

- Not handling allegations of deputy discourtesy reduces the burden on board volunteers to hear many more cases. At the same time, this restriction results in a lack of citizen oversight of these types of incidents.

- Having an attorney present at all hearings provides for instant legal advice, without which there might be additional continuances of cases.

CRB's Web address is *www.qualitywebs.net/crb*.
For further information, contact:

Paul McQuilkin, Ph.D.
Chairperson
Orange County Citizen Review Board
55 West Pineloch Avenue
Orlando, FL 32806
407–823–2821

Capt. Melvin Sears
Administrative Coordinator
Orange County Citizen Review Board
55 West Pineloch Avenue
Orlando, FL 32806
407–858–4797

The Portland, Oregon, Police Internal Investigations Auditing Committee: A City Council, Citizen Advisers, and a Professional Examiner Share Oversight Responsibilities

Background

When the police arrested a number of gay persons in a park in 1993 for alleged sexual activity in public, some neighbors and the arrested individuals complained that the police had used excessive force and had singled out homosexuals for special enforcement. The mayoral candidate promised to look into the problem. As a result, the city auditor prepared an audit of both the Portland Police Bureau's IA unit and the existing Police Internal Investigations Auditing Committee (PIIAC) that had been created in 1982. A local Copwatch organization also submitted a proposal for strengthening citizen oversight. As a result of these efforts, in 1994 the mayor proposed, and the city council approved, changes to the city code that strengthened PIIAC's authority and provided for the appointment of an auditor.

As shown in exhibit 2–11, Portland's oversight structure includes three components: the four-member city council, citizen advisers, and a professional examiner.

- Technically, the city council itself (including the mayor) is the Police Internal Investigations Auditing Committee, although most people loosely refer to the entire oversight procedure as PIIAC. By ordinance, the committee is required to:

    — Assist the police bureau in maintaining community credibility in its internal affairs investigations by issuing public reports on the process.

    — Provide a discretionary review process for complainants who are dissatisfied with an IA investigation.

- The ordinance allows the committee to "utilize Citizen Advisors consisting of 13 persons to assist in performing its duties and responsibilities." Advisers:

    — Hear appeals as a group at monthly meetings from citizens dissatisfied with police internal affairs investigations of their complaints.

    — Review all closed IA cases involving use of force.

    — Individually conduct random audits of IA investigations.

- An examiner, hired by the mayor, coordinates the work of the committee and citizen advisers and conducts much of the auditing herself.

In 1997, citizen advisers processed 21 appeals. The advisers or the auditor monitored 98 cases.

THUMBNAIL SKETCH: PORTLAND

Model: citizens hear appeals (type 3) and audit IA process (type 4)

Jurisdiction: Portland, Oregon

Population: 480,824

Government: strong mayor, city council

Appointment of chief: mayor appoints and can fire

Sworn officers: 1,004

Oversight funding: $43,000

Paid oversight staff: one full time

Appointed by council members and neighborhood coalitions, 13 "citizen advisers" hear appeals from citizens dissatisfied with police investigations of their complaints, review all closed cases involving the use of force, and conduct random audits of IA investigations. The city council, meeting as the Police Internal Investigations Auditing Committee (PIIAC), hears appeals from citizens who are dissatisfied with the police department's investigation of their complaints. A professional examiner coordinates the work of PIIAC and the citizen advisers and conducts many of the audits herself. The examiner and citizen advisers also provide the chief with policy recommendations.

Exhibit 2–11. Steps in the Portland Audit Procedure

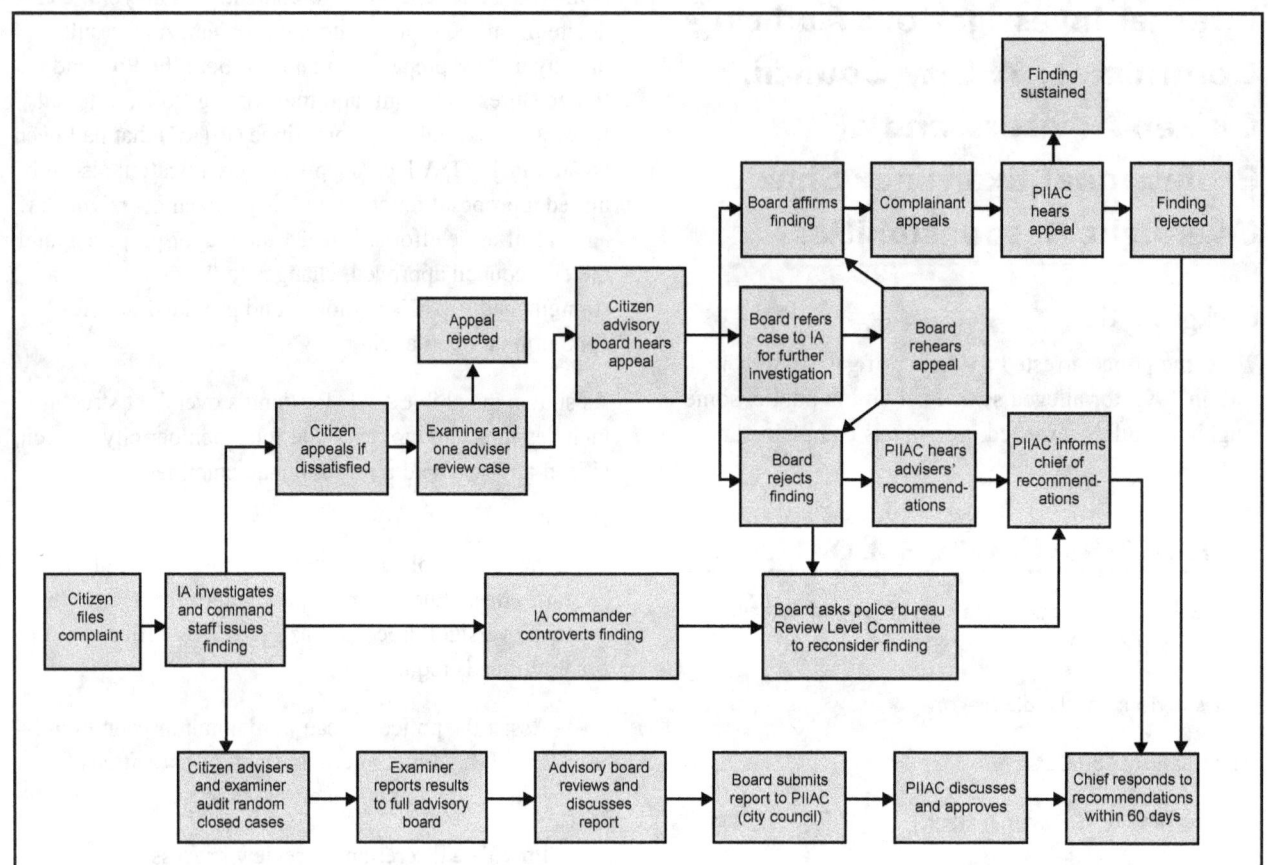

Citizen appeals of IA findings

Citizens may appeal a complaint finding within 30 days after IA has completed its investigation. If the complainant calls the examiner for a hearing, the examiner sends the person an appeal application, schedules the hearing for a future citizen advisory meeting, and arranges for the police bureau to send her the investigation file, which she distributes to advisers to review before their next meeting.

The examiner and an adviser of her choosing go separately to IA to review the case file and confirm each other's assessment of the investigation. The examiner then prepares a report that includes a summary and analysis of the case, a critique of the investigation process, and recommendations for how the case should be handled. She distributes the report to all the advisers and IA to review. At times, she discusses the findings or the investigation process with the IA captain before the advisers meet.

At its next monthly meeting, the full volunteer citizens advisory board can deny the request for review. If the board accepts the appeal, it conducts a formal hearing. Subject officers may attend but usually do not. However, officers' names are not used—they are referred to as "officer A" and "officer B." If he wants to know what transpires, the police union president attends. Someone from IA is present to explain how it investigated the case.

Although the advisers have read the full report before the meeting, the citizen adviser who reviewed the case gives a brief oral case summary to the other advisers. The chairperson then asks the complainant, "Please tell us what you would like us to know about this case." The complainant can not present new evidence because the hearing is an audit, not an investigation. Advisers may question the complainant, the subject officer (if present), and any witnesses who have come. Advisers discuss the case in public and vote to do one of the following:

- Affirm the police bureau's finding.

- Refer the case to IA for further investigation.

- Recommend that PIIAC (i.e., the city council) inform the chief in writing that the finding does not support IA's determination.

- Refer the finding for reconsideration to the police bureau's Review Level Committee (consisting of branch managers, the accused officer's manager, the PIIAC auditor, one citizen adviser, and nonvoting representatives from the police bureau and the city).

On the two or three occasions a year when the advisory board asks the Review Level Committee to reconsider a finding, the examiner and the adviser who reviewed the case participate in the meeting. On occasion, the Review Level Committee has agreed to change an exonerated finding to one of insufficient facts. On one occasion, the chief overruled the Review Level Committee's recommendation and sided with the PIIAC advisers to sustain a complaint.

Advisory meetings last about 90 minutes, with 15–20 minutes devoted to each appeal. After each monthly meeting, the examiner drafts a report for PIIAC summarizing the advisory board's conclusions regarding each appeal.

Audits

Five citizen advisers volunteer to be on a PIIAC monitoring subcommittee. Subcommittee members look at cases chosen at random by the examiner to determine trends in quality, timeliness, and accuracy of the police bureau's IA procedures and investigations.

Internal affairs sends the examiner all closed cases each month—30–40 cases. The examiner assigns the cases to monitoring subcommittee members with a worksheet to guide their review, but she does most of the reviews herself, including all cases that involve the alleged use of excessive force or discrimination and all cases IA sustains.

The examiner or the assigned subcommittee member goes individually to the police bureau to review the entire file for each case in a private room. One of them completes the worksheet with pertinent information about the case. Officers' and complainants' names are not included in the reports. The examiner and adviser spend 2–4 hours examining how thoroughly and fairly the investigation was conducted and whether the finding is solidly supported.

At each monthly monitoring subcommittee meeting, which is open to the public and the press, advisers and the examiner discuss trends they may have spotted in their investigations. (See "Troublesome Trends Revealed by Monitoring Cases.") Based on the audit results, the examiner develops a draft quarterly report, with subcommittee members' help, highlighting shortcomings in the investigations, abuse trends, and recommended policy or training changes (see "Other activities" on page 45). At the next monthly subcommittee meeting, subcommittee members review the report—and pertinent statutes and

TROUBLESOME TRENDS REVEALED BY MONITORING CASES

In 1998, the examiner—Lisa Botsko (at that time)—and citizen advisers noticed that several complainants reported that, when they asked police officers for their badge numbers, the officers would reply, "I don't have a badge number." Technically, this was accurate. However, officers do have identification numbers. After Botsko shared this concern with the police bureau, the chief clarified the pertinent general order to require officers to interpret requests for their badge number as a request for their identification number.

Botsko and the advisers also noticed that a number of incident reports referred to officers' use of a "distraction blow" without explaining its purpose. After inquiring about the behavior, Botsko learned that the police bureau training department taught the distraction *principle* (e.g., pushing the driver's head while prying his or her hands off the steering wheel)—but not a *blow*—as a means of diverting someone's attention. Indeed, the bureau considers a blow to be a use of force that requires explanation in the incident report. It turned out that some officers had learned the distraction blow technique at the State training academy. As a result, the bureau agreed to explain during inservice training that officers always have to explain in their reports why they struck someone and refrain from using incorrect terminology.

general orders that the examiner prepares for them—before the examiner submits it to the full citizens advisory board at a public session for review and approval. The examiner submits the approved report to PIIAC.

During her first year in 1994–95, Lisa Botsko, the examiner, used to send 60 percent of cases back to IA for additional work; by 1998, this had declined to 20 percent because "IA had figured out what I was looking for." Botsko heard investigators saying, "Be careful, or PIIAC will send it back." (See "Auditors Have Identified Problems With IA's Investigations.") Internal affairs also improved its reports because the bureau improved its training and guidelines for IA investigators.

PIIAC (city council and mayor)

The city council conducts PIIAC business once or twice a month during its regular weekly meetings. The council may piggyback other council work onto the PIIAC agenda; at other times, council members may not meet as PIIAC for 2 or 3 months because no appeals reach them. The examiner schedules the meetings and the mayor chairs them.

Typically, PIIAC hears one or two appeals a month. Sessions are open to the public and are televised by a local television cable channel. The complainant may come to the meeting, and someone is present from IA to answer questions. The committee has the power to compel attendance, testimony, and the production of documents and can administer oaths.

The adviser and auditor present each case, and the complainant comments. While the subject officer may sit in with the other members of the audience, he or she is not questioned because the auditor listens in advance to the taped IA interviews and, as needed, has already requested IA to ask any questions of the officer she felt were omitted. Commissioners ask questions throughout. Each commissioner then comments on the case and votes in public. A majority rules. The committee informs the chief in writing of one of the following:

- No additional investigation is warranted.

- IA should reopen the case to conduct additional investigation and report its findings to PIIAC.

- The finding should be changed (see "When PIIAC and IA Disagree on a Finding").

AUDITORS HAVE IDENTIFIED PROBLEMS WITH IA'S INVESTIGATIONS

The types of problems Lisa Botsko, the first examiner, and citizen advisers found in the past with some IA investigations have included:

- Interviewing only officers and no neutral witnesses.

- Neglecting to interview one or more important witnesses.

- Not taking photographs at the scene.

According to Botsko, leading questions asked by IA investigators remained a problem—for example, asking, "Was the subject being deliberately provocative and antagonistic to you?" instead of asking, "How was the subject behaving toward you?" On one occasion, IA investigated two officers who had arrested a juvenile for a sex crime without contacting the boy's parents before removing him from school. On the audiotape of the interviews, the IA investigator examined the parents "under a microscope," but not the officers—for example, challenging the parents' statements but not the officers'. The investigator asked a civilian witness, "What do you mean the officer was screaming?" but did not ask the officer to describe his own behavior.

Botsko and the auditors also have criticized IA and precinct sergeants for not following consistent procedures in collecting evidence regarding citizen complaints, writing reports, and including documentation in the case file. When Botsko reported to IA in 1997 that the precinct sergeants were not producing consistent reviews, the police bureau agreed to implement annual training for sergeants on how to prepare misconduct reports.

WHEN PIIAC AND IA DISAGREE ON A FINDING

PIIAC disagreed with an IA finding three times in 1997. In two cases, the chief disagreed with PIIAC and agreed with IA that there had been no officer misconduct. In one of those two cases, PIIAC voted 4 to 1 to sustain an allegation of misuse of position against an officer who wrote a police report documenting that his neighbors' unsupervised children were making noise on a trampoline late one night after the officer had tried to resolve the problem by talking with the parents. The report suggested that the State's Child Services Division, which investigates child abuse cases, become involved. The chief supported IA's exoneration of the officer. In the third case, PIIAC decided that IA was incorrect in deciding that an officer had not violated bureau policy by removing a child from school without notifying the child's parents. The officers felt that informing the principal was adequate notification. The chief sided with PIIAC.

Other activities

Based on its audits, the examiner recommends policy and procedure changes to the police bureau in her quarterly reports that the city council, acting as PIIAC, votes to adopt. The chief must respond to the report in writing within 60 days. The response must indicate what policy or procedural changes within IA, if any, he has instituted as a result of the report. If the chief does not respond within 60 days, the examiner sends an e-mail reminding him or telephones the IA commander. If the chief still fails to respond, the city council can consider the matter. Chapter 3 presents illustrative policy recommendations PIIAC has made that the bureau has adopted.

Staffing and budget

Each of the four city council members appoints one adviser; the police commissioner, who also is the mayor, appoints two advisers; and each of seven neighborhood coalitions chartered by the city recommends an adviser to the city council for appointment. Advisers serve for 2 years, subject to reappointment.

The mayor appoints and funds the examiner, who spends full time on oversight activities. The examiner's salary is $43,000. The mayor's office also pays for oversight-related duplication, telephone, and secretarial costs.

Distinctive features

The most unusual features of Portland's oversight system are, first, the use of citizen advisers to review completed internal affairs investigations at the police station and, second, the city council's role in hearing citizen appeals.

- By trying to ensure that IA investigations are done properly, the auditor's approach may eliminate the need for independent professionals to investigate citizen complaints. This approach may reduce the costs of citizen oversight.

- Because PIIAC does not accept citizen complaints, some individuals may not report allegations of police misconduct because they may be afraid to take their complaints to the police bureau. Citizens do have the option of filling out complaint forms at the neighborhood coalitions represented among the citizen advisers, which then forward the forms to IA.

- PIIAC examines only completed cases. As a result, PIIAC cannot shape the conduct of individual investigations while they are in progress. However, through its audits, PIIAC may be able to motivate investigators to do a better job overall. By not investigating cases, the oversight procedure may receive better cooperation from the police.

- Citizen advisers are not professional auditors. As a result, they may not possess, or may need time to learn, the skills needed to conduct a competent audit.

- Because a majority of citizen advisers are chosen by neighborhood associations, citizens may be more likely to feel they are well represented in the oversight process than if advisers were chosen by city officials.

- The system does not require police officers to participate in the audit process.

- Because PIIAC and advisory board meetings are public, and because PIIAC must publish periodic reports, the media have an opportunity to focus on police misconduct, examine how the police bureau conducts its internal affairs investigations, and publicize what they learn.

- PIIAC commissioners—even though they make up the city council—do not have the power to overrule the chief's decision to sustain or not sustain complaints. Police administrators are likely to feel it is important that the ultimate decision remain with the department. Complainants may feel frustrated that elected officials do not have the final say in their cases.

- Because the auditor works for the mayor, the chief executive is free to increase or decrease the hours she devotes to PIIAC.

For further information, contact:

Examiner
Police Internal Investigations Auditing Committee
1221 Southwest Fourth Avenue, Suite 340
Portland, OR 97204–1995
503–823–4126

Bret Smith
Commander, IA Unit
Portland Police Bureau
1111 Southwest Second Street
Portland, OR 97204
503–823–0236

The Rochester, New York, Civilian Review Board: Trained Mediators Review Citizen Complaints

Background

In 1976, after community groups expressed serious concern when police officers killed a woman who was brandishing a knife, the mayor appointed a commission to explore how to improve police-community relations and reduce the use of excessive force. One of the panel's recommendations was a citizen review process. As a result, the city council approved legislation establishing a Complaint Investigation Committee, consisting of

> **THUMBNAIL SKETCH: ROCHESTER**
>
> Model: citizens review cases (type 2)
>
> Jurisdiction: Rochester, New York
>
> Population: 221,594
>
> Government: strong mayor, city council
>
> Appointment of chief: mayor nominates, council approves, mayor may remove
>
> Sworn officers: 685
>
> Oversight funding: $128,069
>
> Oversight staff: one full time, three part time
>
> The Rochester City Council contracts with a local dispute resolution center to set up three-member panels of trained, certified mediators to review internal affairs cases. The panels establish findings that the chief considers along with IA's findings in imposing discipline. The panels also may recommend change (related to the cases it reviews) in department policies, training, and IA investigation procedures. In a separate process, the dispute resolution center conducts about 10 formal citizen-police conciliations each year.

review panels with two command police officers and one citizen that met at police headquarters to review completed IA investigations. In 1984, the council changed the composition of the panels to include two civilians and two command officers and established a conciliation process. In 1992, the council renamed the committee the Civilian Review Board (CRB), excluded any police representation, and moved the reviews to the city hall.

The city council contracts with the Center for Dispute Settlement to train and provide the panelists and arrange for the reviews. Founded in 1973 by the American Arbitration Association, the center is the third oldest not-for-profit dispute resolution organization in the Nation. It offers alternative dispute resolution options to the court system and trains community members to conduct conciliation. The city council chose the Center for Dispute

Settlement to perform the citizen oversight function because it appeared to be the most capable organization in the city for conducting an impartial review of police behavior. Although most of the center's funding comes from the New York State Unified Court System, CRB's budget is a line item in the police department's budget.

In 1997, the police department submitted 26 completed cases out of 131 for CRB review, or 20 percent. The 26 cases involved 80 allegations. For the first 9 months of 1998, CRB reviewed 58 cases involving 141 allegations. CRB sustained 23 percent of cases in 1997 but only 7 percent during the first 9 months of 1998.

Procedure

Exhibit 2–12 shows CRB's review process, which is discussed in detail in the following section.

Intake

Citizens may file complaints by mail or in person at the Center for Dispute Settlement office as well as at city hall or police headquarters. The center received nine complaints in 1997, most referred by the mayor's office, and five complaints during the first 9 months of 1998. After screening to make sure the case has merit or is suitable for conciliation (see following), the center forwards the complaints to the police department's internal affairs unit for investigation. The vast majority of complainants file directly with the police department.

IA sends for CRB review all investigations of cases that involve:

- Charges of use of excessive force.
- Conduct that, if proven, would constitute a crime.
- Other matters the chief chooses to refer to CRB.

During or after completion of each internal affairs investigation, IA calls Todd Samolis, the CRB coordinator, to notify him to set up and schedule a panel to review the completed case.

Hearings

Each CRB panel consists of three volunteers selected from a pool of 15–20 individuals who are certified mediators, have attended a shortened version of a police academy (see below), and receive special training to function as panelists. One of the three panelists is a chairperson who facilitates the review. CRB held as few as two panels in January 1998 and as many as 13 in June; the modal number (occurring in each of 5 months in 1998) was 7.

EXHIBIT 2–12. ROCHESTER CITIZEN OVERSIGHT PROCESS

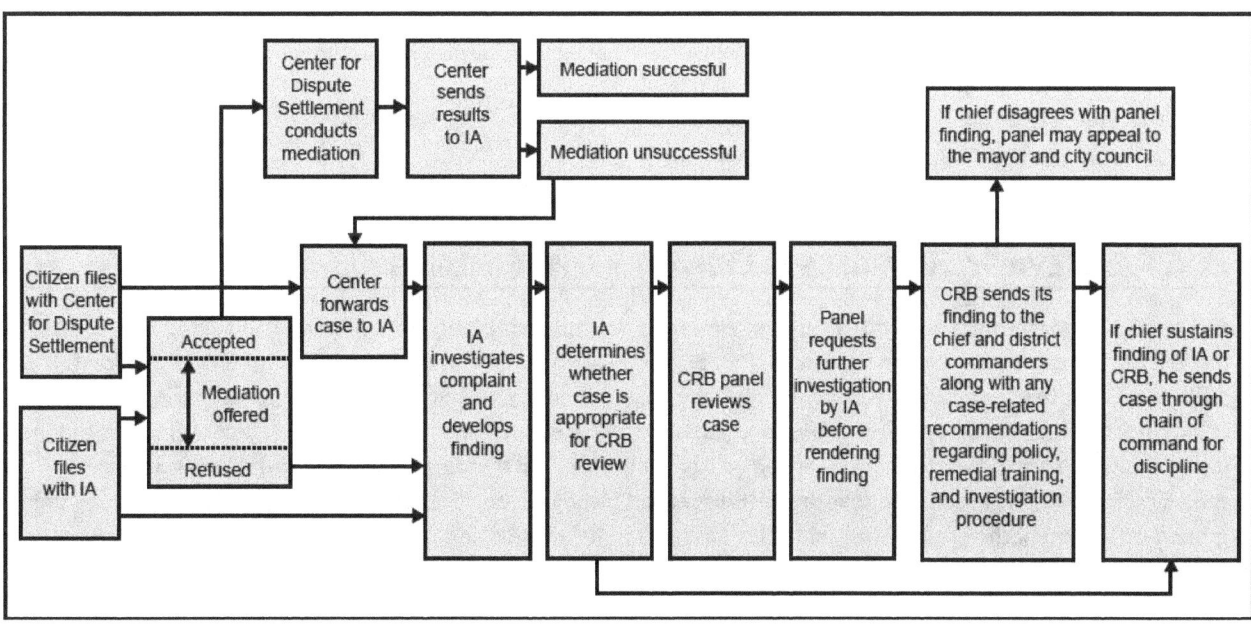

Reviews are held during the day in a soundproof, locked room in the city hall basement where the CRB files are kept under lock and key. The investigating sergeant for a given complaint brings the key to the room and unlocks it for the panelists before the review.

The sergeant begins the session with a 3–5 minute summary of the case, distributes copies of the case file (including pertinent department policies and procedures), and leaves. However, the investigator leaves a pager number so that panelists can call with questions during their deliberations. The CRB legislation also requires the department to make available an officer with the rank of captain or higher, neither from IA nor a commander of the officer involved in the case, to answer questions related to department policy and procedure. For example, panelists once called the designated captain to ask whether it was department policy that officers take all subjects sprayed with Mace to the hospital; he informed them that officers have the discretion to take them to the police station basement to wash out their eyes.

Panelists do not have access to the IA investigator's case file in advance. Instead, they review the file after the investigator has left. After the panelists have completed their silent review, the chair introduces the allegations one by one. Each member gives his or her recommended finding and rationale. Questions and discussion follow.

Findings

On occasion, panelists ask the investigating sergeant to conduct additional investigation, such as interviewing a new witness or reinterviewing an existing witness. If the panel is still unsatisfied with the quality of the investigation, it can appeal, in sequence, to the IA commander, the chief, the mayor, and the city council. The city council, with its full subpoena power, can itself interview witnesses and request documents. Panels have never needed to go beyond the IA sergeant to request additional investigation. (See "A CRB Review Reverses a Department Finding.")

At the end of the discussion (if they have not requested any additional investigation), the chairperson tape records the panel's finding and justification. As can IA investigators, panelists may choose among four findings: unfounded, exonerated, unprovable, and sustained. Panels make their determination based on a preponderance of the evidence. Although panelists do not vote, they disagreed on only 5 of the 141 allegations they reviewed during the first 9 months of 1998. When not unanimous, the dissenting panelist may read his or her finding into the tape along with the rationale for dissenting. After the taping, the chairperson opens an envelope the investigating sergeant left that contains IA's findings. The panels' findings are consistent with the IA findings about 95 percent of the time.

When the session is over, the chairperson telephones the investigator, who retrieves the tape and written report. (City hall is a 5-minute walk from police headquarters.) The IA unit sends CRB's findings, along with its own findings, to the subject officer's division commander, the deputy chief, and the chief for review. If the chief sustains the finding, the case goes through the chain of command for penalty recommendations, starting with the officer's sergeant and ending with the chief, who makes the ultimate disciplinary determination.

In 1997, the chief disagreed with 6 of the 80 panel findings. In all but one of the six, he increased the severity of

A CRB Review Reverses a Department Finding

Police officers got into a tussle with a suspect. An officer hit the man in the face and then handcuffed him. The man filed a complaint alleging improper use of force. An IA investigation cleared the officer of any wrongdoing. CRB concluded the complainant was right. When the case came back to the department and went through the chain of command, the deputy chief said he agreed with the CRB panel and asked IA to do additional investigation. Based on its additional investigation, IA ended up agreeing with CRB's finding. According to Lt. James Sheppard, the IA commander, "It turned out that the CRB panel had picked up on the fact that the man was lying flat on the ground on his stomach with his arms under his chest when he was hit, and passive noncompliance does not justify hitting a person."

the finding—in four cases he changed it from unfounded to unprovable. In each case, there were no independent witnesses to verify the account of the incident. When the chief disagrees with a CRB finding, the coordinator can take the disagreement to the mayor or the city council, but he has never done so.

Other responsibilities

CRB also suggests policy changes, remedial training, and changes in IA investigation procedures to the department, and the Center for Dispute Settlement mediates selected citizen complaints.

Policy and other recommendations

CRB can make recommendations to the chief regarding revisions to police policies and procedures relevant to a given case. Although CRB does not recommend discipline, panelists may recommend case-related remedial training.

> A mother and daughter filed a complaint because they felt they were being treated as suspects when they called the police to disperse some gang members who would not leave their porch. The mother and daughter objected so strongly to the officers' attitude that the officers ended up arresting the two women. A CRB panel exonerated the officers but recommended they be retrained in interviewing and conflict resolution skills. The chief ordered the retraining.

Todd Samolis, the CRB coordinator, meets with his IA counterpart every 3 months to go over each case to learn whether any policy, training, and investigation procedure changes that panels may have recommended were implemented. The chief sends CRB new or revised general orders that result from a panel recommendation.

Conciliation

In 1984, a city council member suggested the Center for Dispute Settlement provide a conciliation option in an effort to help build positive relations between officers and citizens. Cases involving allegations of excessive use of force are not eligible for conciliation.

Depending on where the complainant files the case, either Todd Samolis or an IA investigator may ask the person if he or she would find conciliation an acceptable alternative to an IA investigation and CRB review if the officer also agreed to the procedure. Most complainants offered the option to conciliate accept. When an officer's supervisor presents an officer with the option, about half comply.

For conciliating citizen complaints against the police, Samolis chooses one of the center's mediators who have participated in a 1-day extra training session on police conciliation. Conciliations are confidential. The parties sign no written agreement. Instead, the mediator indicates in the case file whether in the complainant's judgment the matter was resolved or unresolved.

If the matter is resolved, Samolis sends a letter indicating closure to IA. Internal affairs does not investigate the complaint, and the case is closed. Samolis notifies IA if the matter is not resolved and the complainant wishes to have the complaint investigated. In 1997, three out of the four conciliations were successful. Of five conciliations conducted from January through September 1998, two were resolved, one was unresolved, for one the complainant did not appear, and for one the officer did not appear.

Staffing and budget

CRB's activities are administered by Todd Samolis, the full-time coordinator; by the half-time support of the Center for Dispute Settlement director of special programs; and by the quarter-time support of the center's director of training services. Candidates for panelist positions must first attend the Center for Dispute Settlement's 25-hour principles of mediation course that provides State mediation certification. The course includes extensive training in how to be impartial. Candidates then serve an apprenticeship that involves observing regular mediators in two or three sessions, co-mediating two or three sessions with an experienced mediator, and conducting an observed pass/fail solo mediation session. Finally, candidates attend a 2-week, 48-hour condensed version of a police academy run by the police department that includes using sidearms with a "Shoot/Don't Shoot" simulator, handcuffing, and explanations of department policies and procedures.

CRB administrators nominate experienced panelists who have demonstrated exemplary ability as permanent chairpersons. The mayor approves their selection. CRB arranges for one of the chairpersons to run each panel before contacting two regular panelists to complete the panel.

The Center for Dispute Settlement's fiscal year 1998–99 budget for CRB and conciliation was $128,069 (see exhibit 2–13). The budget includes $17,000 for panel member and mediator stipends.

Distinctive features

Rochester's use of trained mediators to review cases is the oversight procedure's most innovative feature.

- According to Todd Samolis, the CRB coordinator, "Training as mediators goes to the essence of objectivity, including promoting listening skills, asking

EXHIBIT 2–13. CENTER FOR DISPUTE SETTLEMENT CRB AND CONCILIATION BUDGET FOR FISCAL YEAR 1998–99

Personnel Costs % of Full Time	Full-Time Salary	Position	Budget
55	$33,114	director of special programs	$18,213
25	$31,353	director of training services	7,838
100	$21,011	program coordinator	21,011
60	$15,288	program assistance—clerical	9,173
Total salaries			**56,235**
FICA (.0765)			4,302
Fringe (.1035)			5,820
Total personnel costs			**66,357**

Other costs	
Stipends:	
Conciliation/mediation	
$35/2 hours/any part thereof	500
CRB reviews	
$35/2 hours/any part thereof	15,800
(70 cases × 2 people) = $9,800)	
$50/2 hours/any part thereof	
(60 cases × 1 person) = $6,800)	
Quarterly CRB chair meetings	
$35 per meeting	700
(Based on four meetings with five CRB chairpersons per quarter)	
Training and outreach	5,000
Training inservice (four sessions @ $150)	600
Printing	500
Postage	400
Space ($685/mo.)	8,220
Telephone	720
Supplies	550
Miscellaneous service (database management system)	500
Equipment rental ($65/mo.)	780
Parking/mileage ($12.50/mo.)	150
Insurance	200
Conference	4,500
Total other costs	**39,120**
Subtotal	105,477
Administrative overhead 22%	22,592
Total projected budget 1998–99	**$128,069**

probing and open-ended questions, developing a rationale for each position taken, and looking at all sides of a problem." However, if CRB's parent organization were not a dispute resolution center, arranging to train board members as mediators might be expensive.

- Panelists hold the position for life (subject to proper behavior). On the one hand, their longevity provides them with considerable experience reviewing cases that may enable them to act efficiently and objectively. On the other hand, according to Robert Duffy, the chief of police, "Some of them may sort of 'settle in' and lose the fresh perspective citizens can bring to police work." In addition, rotating panelists would enable CRB and police to educate more community members to the nature of police work.

- By designating permanent chairpersons, usually from among long-time panelists, only the most qualified and experienced panelists facilitate the reviews.

- Cases are reviewed relatively quickly. (The city council deliberately chose a system that would avoid the delays it found existed in some other jurisdictions.) According to the city council resolution establishing the board, CRB has to review cases within 2 weeks of IA's notification that its investigation is complete.

- Panels are anonymous and not open to the public. As a result, panelists are not under pressure to skew their decisions in response to the demands of public or police interest groups. However, the public may lack confidence in CRB's objectivity since citizens are not privy to the review process.

- Panelists do not have an opportunity to review IA case files before the panel meets. This results in extra time being taken during the meeting while panelists review the files and may create pressure to review the materials less thoroughly than if panelists could review them at home at their leisure before the meeting. Panelists also do not have the opportunity to ponder the cases in advance of the meetings. However, by not distributing any IA case files outside the meeting room, the police department is assured they will never be made public, for example, by getting lost.

- By not handling allegations of police discourtesy and other less serious complaints (unless they are part of a serious complaint), CRB can focus on more important cases. However, the board might be able to handle many less serious cases through mediation with greater satisfaction to complainants and more objectivity than the police department may be able to achieve.

For further information, contact:

Todd Samolis
Coordinator of Special Projects
Civilian Review Board
300 State Street, Suite 301
Rochester, NY 14614
716–546–5110

Robert Duffy
Chief of Police
Rochester Police Department
City Public Safety Building
Civic Center Plaza
Rochester, NY 14614
716–428–7033

The St. Paul Police Civilian Internal Affairs Review Commission: A Police-Managed Board Recommends Discipline

Background

Because of complaints about police misconduct, and in the aftermath of the Rodney King beating in Los Angeles, William Finney, the St. Paul police chief, urged the city council to establish a commission to look into forming a civilian oversight procedure. The resulting Police Civilian Internal Affairs Review Commission began operation in December 1993. Located in the fire department wing of the public safety building, the commission is operated by the police department.

The commission met 12 times in 1997 to review 71 cases involving 149 allegations (73 of them involved the alleged use of excessive force). The commission's findings were as follows:

- Unfounded: 53 (36 percent).

- Exonerated: 32 (22 percent).

Chapter 2: Case Studies of Nine Oversight Procedures

Thumbnail Sketch: St. Paul

Model: citizens review cases (type 2)

Jurisdiction: St. Paul, Minnesota

Population: 259,606

Government: strong mayor, city council

Appointment of chief: mayor nominates, council approves, mayor may remove only with council approval

Sworn officers: 581

Oversight funding: $37,160

Oversight staff: one part time

A seven-person commission that is part of and funded by the St. Paul Police Department meets monthly to review cases investigated and decided by the department's internal affairs unit. The commissioners, two of whom are St. Paul police officers, make their own findings and, in sustained cases, recommend discipline to the chief. The IA unit makes no disciplinary recommendations. The chief is free to disregard the commission's disciplinary recommendations but not its findings.

- Not sustained: 41 (28 percent).
- Sustained: 23 (15 percent).

The commission also reviewed 24 cases of discharge of firearms and found them all to be justified. The commission found a policy failure in two cases.

The review process

Exhibit 2–14 displays the citizen review procedure in St. Paul.

Intake

The St. Paul administrative code requires the commission to review all completed IA investigations related to:

- Alleged acts of excessive force.
- Use of firearms (regardless of whether there has been a citizen complaint—see "The Review Commission Hears All Discharge-of-Firearms Incidents").
- Discrimination.
- Poor public relations.
- Other complaints the chief or mayor chooses to refer to the commission. (The chief sometimes refers internal complaints, particularly sexual harassment cases.)

The Review Commission Hears All Discharge-of-Firearms Incidents

By statute, the review commission hears all cases in which an officer discharged a firearm, regardless of whether a citizen filed a complaint. Most of the cases involve euthanizing injured animals, especially deer. Others involve accidental discharge.

> An officer had drawn her sidearm while searching a warehouse for a reported burglar. She had left the building to climb a grassy hill next to the warehouse to continue the search when the man ran out of the building. The officer yelled at the man to stop. She then slipped on the wet grass and fired her gun accidentally. The man, thinking she had fired a warning shot, stopped running and was arrested by another officer.
>
> The officer reported the entire incident fully, but the department forbids the accidental discharge of weapons. As a result, IA found her guilty of misconduct, and the commission did, too. The commission recommended she receive additional firearms training.

EXHIBIT 2–14. ST. PAUL CITIZEN REVIEW PROCESS

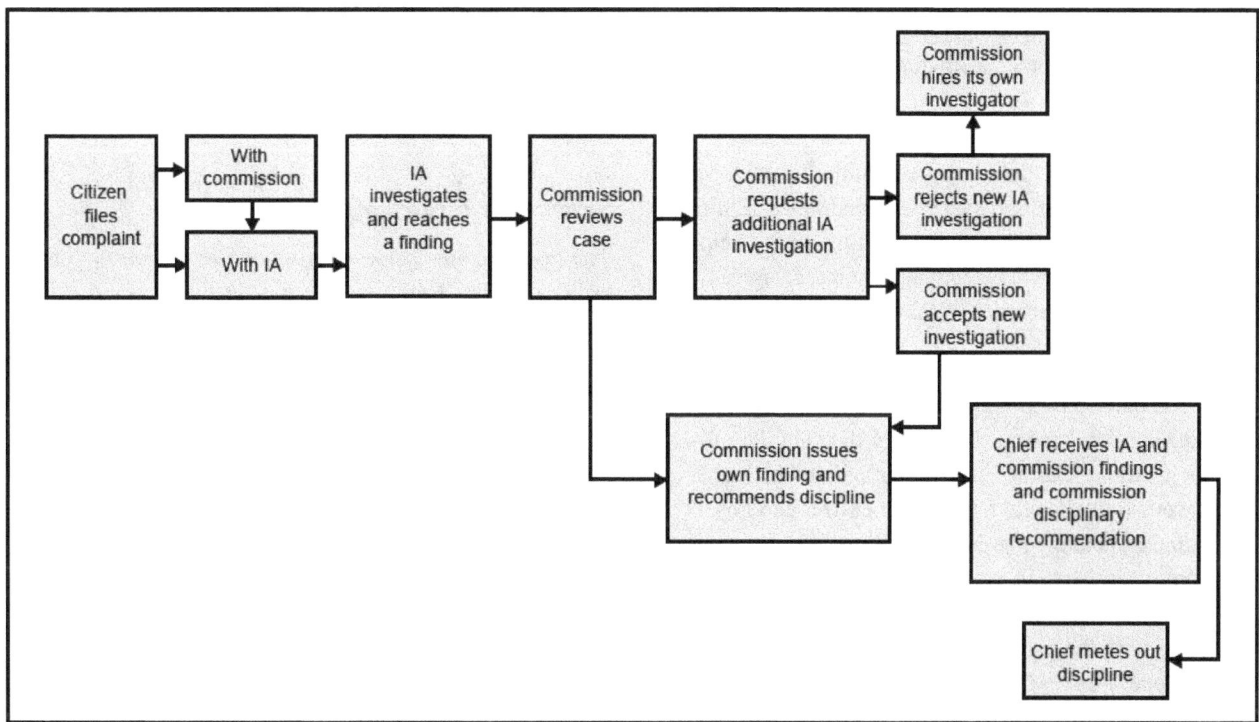

While about three-quarters of complainants contact the IA unit to file a complaint, some contact the commission. The review commission coordinator takes basic information about the complaint and refers the complainant to the IA unit. The unit investigates serious allegations of misconduct itself but refers minor problems down the chain of command to the subject officers' supervisors for settlement. The IA commander reviews these settlements but does not send them to the commission for review.

Case review
The commission coordinator, a nonsworn police department employee, collects IA's investigative packets 2 weeks before each commission meeting, duplicates them, and hand delivers a copy to each commissioner (some commissioners pick them up in person) 1 week before they meet. The commission meets in the chief's conference room the first Wednesday of every month from 7:00 p.m. to about 10:00 p.m.

Because the Minnesota Data Practices Act gives employees (e.g., police officers) privacy in administrative hearings, only commissioners, the commission coordinator, the IA commander and investigators, and a recording secretary attend. Members hear about seven cases each meeting. The commission may request—and has subpoena power to require—that individuals appear before it.

The IA investigator summarizes the first case and gives his or her recommendation that the allegations be sustained, not sustained, exonerated, or unfounded. The commissioners discuss each case, asking the investigator questions as needed. The chairperson asks for a vote on the first allegation in the case. If a majority sustains an allegation, they discuss what discipline to recommend. Deliberations typically take 15–45 minutes per case. Most decisions are unanimous. When there is disagreement, it is usually regarding the discipline, not the finding. The commissioners may request that IA conduct additional investigation. If they are still dissatisfied, they can hire an independent investigator, although they have never done so.

The IA commander and investigator play no role in helping the commissioners to resolve their differences and may not object to the commission's recommendations. Commissioners may ask the commander what kinds of discipline have been imposed before for the misconduct if it is a new type of wrongdoing. Commissioners have access to the officers' previous disciplinary records and can therefore recommend "progressive discipline"—more serious sanctions for repeat offenders.

After the hearing
The chair sends the chief a memorandum after each hearing with the commissioners' and IA's findings and, if the complaint has been sustained, the commission's disciplinary recommendation. The commission has disagreed with IA's finding in about a half dozen cases in its history.

> A citizen complained that an officer's remark to a block party, "Don't call me unless you call public housing first," meant that, if they did call, the officer would not come. The officer claimed he had never said he would not come if called, and no witnesses claimed he had said he would not come. Although IA had sustained the complaint, the commission exonerated the officer. The chief sided with the commission.
> —Donald Luna

Since the commission's first meeting, William Finney, the current chief, has given it the additional task of recommending disciplinary penalties for sustained cases. Although the chief is not obligated to follow the commission's disciplinary recommendations, Finney estimates that he disagrees with less than 2 percent of the sanctions that the commission recommends. On one occasion, Finney met with the entire commission to explain why he chose to deviate from a recommended discipline. When he has disagreed with the commission, Finney has usually increased its recommended discipline. (See "The Chief Increases the Commission's Recommended Punishment.")

There is no appeal of the commission's and chief's dispositions.

Staffing and budget

The review commission consists of five citizens and two police officers. The mayor nominates the citizen members and the city council approves. The police union's executive board nominates the two sworn members for approval by the membership at a union meeting. Once approved, the chief recommends them to the mayor for appointment. The two sworn police officers receive overtime pay if meetings do not occur during their regular shifts. All commissioners serve 3-year terms, renewable once. The commission elects a chairperson and vice chairperson from among the citizen members to preside over its proceedings.

A coordinator, appointed by the chief in consultation with the commission chairperson, spends half of her time managing the complaint process. (She spends the rest of her time coordinating the citizens' police academy.)

Exhibit 2–15 shows the commission's budget for 1995 (which has remained largely unchanged in subsequent years). As shown, the entire appropriation was $37,160, including half of the director's salary and $18,660 in direct costs. However, because the commission has never exercised its option to hire an independent investigator, the true costs are closer to $27,000 per year.

THE CHIEF INCREASES THE COMMISSION'S RECOMMENDED PUNISHMENT

According to Chief William Finney, "We had an officer who had a 'smart mouth,' but she had never had a sustained complaint. When she finally got a sustained finding from both IA and the commission, the commission recommended supervisory counseling. However, because I knew she had had a history of 'mouthing off,' I suspended her for a day."

Donald Luna, the review commission chair, has a similar story: A number of citizens were playing games with an officer regarding the owner of a car that the officer was trying to have moved: "It's not my car, it's his car; no it's her car." They also began calling him derogatory names. After an hour of this, someone in the crowd said, "Why do you have to be such an a-----e?" The exasperated officer answered, "Well, I guess I'm just an a-----e. Now *move the car*." A minister heard the remark and filed a complaint. The commission recommended supervisory counseling; the chief gave him a 3-day suspension.

Exhibit 2–15. Police Civilian Internal Affairs Review Commission 1995 Budget

Budget Item	Funding Level
Coordinator's salary (1/2 time)	$18,500
Direct costs	18,660
commissioner stipends ($50 per meeting)	6,000
consultants to train new commissioners	700
business cards	60
independent investigator	10,000
office supplies	200
conference attendance by coordinator	1,500
miscellaneous training (e.g., seminars) by coordinator	200
Total	**$37,160**

Distinctive features

The two special features of the St. Paul oversight system are the review commission's inclusion of two active police officers from the city and its mandate to recommend discipline.

- On one hand, some members of the community may not see the commission as capable of being objective because it has two officers as commissioners, is part of the police department, and meets at the police station. As a result, some citizens may not trust the process. On the other hand:

  — Because of who they are and their familiarity with how officers have been trained, the sworn members of the commission tend to be tougher than the civilian members in their findings and recommendations for discipline.

  — Having two sworn officers on the board reduced friction between citizen review advocates and the police union and other officers when the board was being planned and in its subsequent operation.

- There are no disputes over gaining access to IA reports in a timely fashion because the commission is internal to the department.

- Although the chief is not obligated to follow the commission's disciplinary recommendations, the commission's ability to provide disciplinary advice allows the chief to learn how community representatives view each officer's misconduct and to impose punishment, if he so chooses, that reflects these representatives' concerns.

- The St. Paul oversight procedure provides no public forum for individual citizens and organizations to express complaints and concerns about the police department's policies and procedures and officers' behavior.

- Officers are spared having to appear before the commission, but some may feel frustrated that they cannot present their side of the story in person.

For further information, contact:

William Finney
Chief
St. Paul Police Department
100 East 11th Street
St. Paul, MN 55101
612–292–3588

Ruth Siedschlag
Coordinator
Police Civilian Internal Affairs Review Commission
100 East 11th Street
St. Paul, MN 55101
612–292–3583

San Francisco's Office of Citizen Complaints: An Independent Body Investigates Most Citizen Complaints for the Police Department

Background

In 1982, the San Francisco City/County Board of Supervisors put the Office of Citizen Complaints (OCC) on the ballot as a voter initiative after a series of police clubbings of demonstrators led to pressure for a citizen oversight procedure from liberal organizations and historically discriminated-against communities, including the department's own African-American Officers for Justice. A police commission, consisting of five members appointed by the mayor, supervises both OCC and the police department. The commission hires the chief and OCC director. The commission or the mayor may remove

CHAPTER 2: CASE STUDIES OF NINE OVERSIGHT PROCEDURES

> **THUMBNAIL SKETCH: SAN FRANCISCO**
>
> Model: citizens investigate (type 1)
>
> Jurisdiction: San Francisco
>
> Population: 735,315
>
> Government: strong mayor, city/county Board of Supervisors
>
> Appointment of chief: police commission (appointed by the mayor) appoints, commission or mayor may remove
>
> Sworn officers: 2,100
>
> Oversight funding: $2,198,778
>
> Oversight staff: 30 full time
>
> An independent Office of Citizen Complaints (OCC), with 15 full-time investigators, investigates most citizen complaints against the San Francisco Police Department and prepares findings. If the department's internal affairs division agrees with the OCC finding, the case usually receives a chief's hearing at which the assistant chief presides. An OCC attorney prosecutes the case. The assistant chief typically approves the finding and metes out discipline subject to the chief's approval. The police commission holds an administrative trial for cases of alleged serious misconduct at which an OCC attorney again acts as the prosecutor. OCC also provides policy recommendations to the department, arranges for mediation, and provides early warning system data.

the chief. Only the commission may remove the OCC director.

OCC received 1,126 new complaints in 1997. Of 983 cases closed in 1997, OCC sustained one or more allegations in 100 cases, or 10 percent. In 1997, OCC held more than 50 hearings at the chief's level and prosecuted 6 cases before the police commission.

The complaint filing process

Exhibit 2–16 diagrams the civilian oversight process in San Francisco. The following text describes each step in the process.

Intake

The police department's internal affairs division (technically called the Management Control Division) alone investigates complaints brought by officers against each other and incidents involving off-duty officers and nonsworn personnel. Internal affairs and OCC both investigate cases involving use of a firearm. OCC alone investigates cases citizens initiate alleging misconduct—or failure to perform a duty—by on-duty officers.

Internal affairs sergeants offer to help citizens who appear at the police station to fill out the complaint intake form and forward it to OCC, but more than half of these citizens choose to go to OCC (a 15-minute walk from police headquarters) to fill out the form. Complainants also may telephone, mail, or fax their complaint to OCC. Of the 1,126 complaints OCC received in 1997, 43 percent were made in person, 23 percent by phone, 21 percent by mail, 5 percent at the police department, and 6 percent by other means. Organizations filed 24 of the complaints in 1997 (see "Organizations May File Complaints").

When a complainant appears in person, the receptionist asks the person to fill out an intake form. The receptionist calls the intake investigator for the day (each investigator generally does intake 1½ to 2 days a month) or, if he or she is busy or on break, the backup intake investigator (who is the following day's intake investigator). If the citizen telephones to file a complaint, the receptionist may refer the call to an investigator and generally mails the person the intake form to complete and mail back. After OCC has received the form, an investigator telephones the complainant and conducts a telephone interview or arranges to interview the citizen in person. In serious cases, OCC makes an investigator available 24 hours a day, 7 days a week.

The initial interview with the complainant

California State law and the police commission prohibit revealing any information about a complaint to the public unless the case is heard by the police commission (see "Police commission hearings" on page 59). The investigator therefore tells the citizen the complaint will be kept confidential unless the person makes it public. As a result, the investigator cannot locate or interview witnesses by telling people about the complaint unless the complainant agrees to their being told about the complaint.

Because investigators can log onto the police department's computer, they can find the computer-aided dispatch data during the interview to identify which officers were at the scene as well as to download the incident report. Investigators also review the pertinent general

Exhibit 2–16. San Francisco's Oversight Process

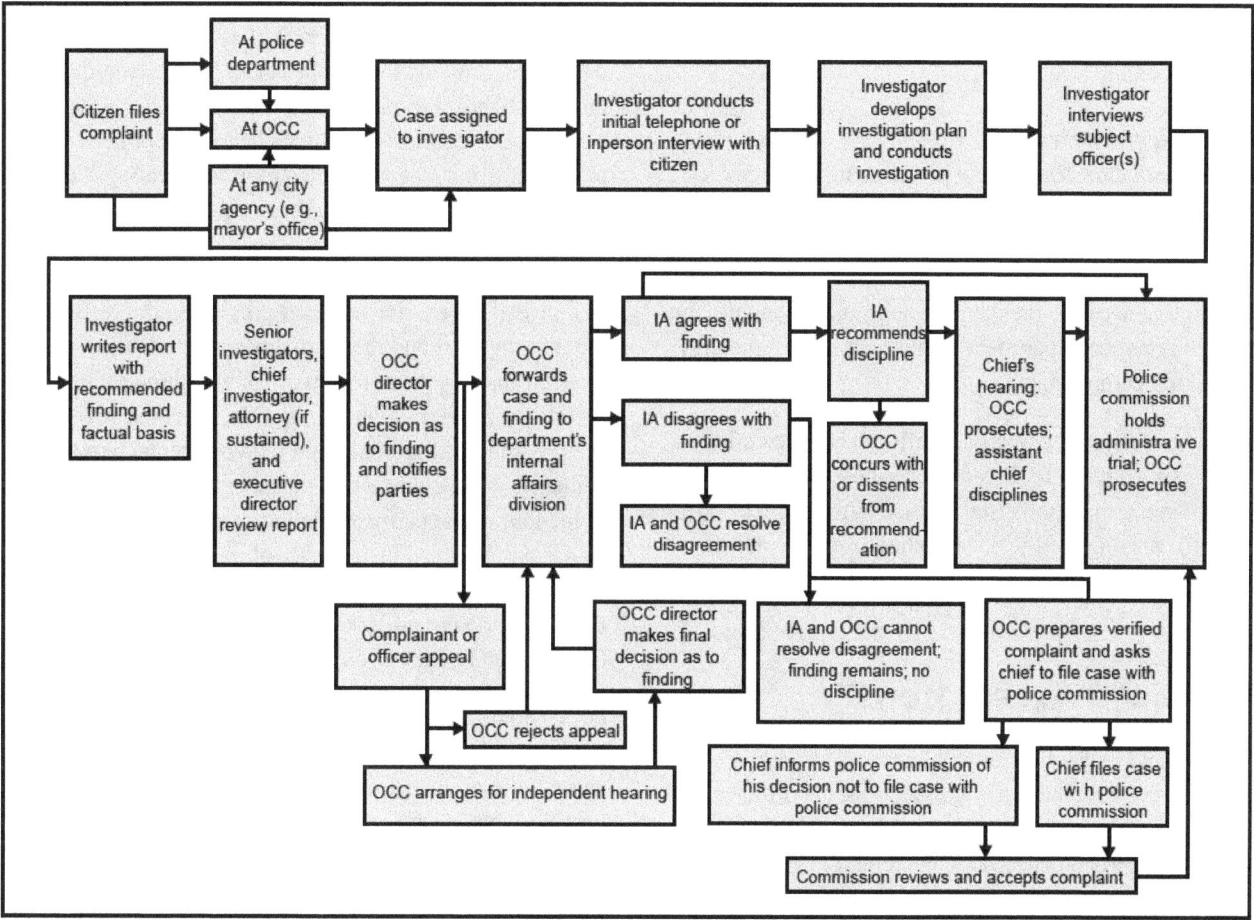

ORGANIZATIONS MAY FILE COMPLAINTS

The San Francisco Bay Area chapter of the National Lawyers Guild provided legal support for a demonstration in 1997 on the anniversary of the Rodney King beating in Los Angeles. After the demonstration, the guild mailed a complaint to OCC alleging that police officers arrested demonstrators who followed instructions to get onto the sidewalk along with demonstrators who refused to move. OCC investigated the incident and notified the guild that it had not sustained the allegation. In the meantime, the guild had received a positive judgment in small claims court for false arrest and was awarded damages. Based on these new developments, the guild asked OCC to reopen the case, but the oversight body denied the request because the director believed that a videotape clearly documented the demonstrators to be in the wrong and that the OCC finding was therefore not in error. In addition, OCC determined that it had investigated the case fully and was given no new evidence that would merit granting the appeal.

orders and department bulletins to determine whether the officers may have violated department policy.

Investigators tape record all interviews except when citizen informants refuse to serve as witnesses. The typical interview lasts 15 minutes to 1 hour. OCC gives the complainant a copy of the complaint form and sends another copy to each named officer providing notification of the allegations as required by State law. OCC sends a copy to the officer's commanding officer and to internal affairs. (See "Added Allegations.")

The officer interview

The investigator develops an investigation plan that includes interviewing the involved officer(s) and any witnesses and reviewing available written documents. The plan may also include collecting evidence, such as visiting the scene of the incident, photographing vehicles, and using the police department's photo lab to take pictures of injured complainants.

Investigators interview officers after they have collected sufficient evidence to determine the best questions to ask. Investigators may not sustain a case without interviewing the officer in person. It is a violation of police department general orders for subject officers to refuse to attend and answer questions at an OCC interview. If an officer ignores the request after investigation by internal affairs, the department generally handles a first violation with an admonishment, the second with a reprimand, and the third with a 1-day suspension.

Investigators generally prepare questions in advance and follow written guidelines in their initial questioning of subject officers. A union representative often comes to the interview with the officer. During the interview, some union representatives raise objections for the record, which the investigator has no authority to rule on. Objections may be resolved later at a chief's hearing or police commission hearing (see "Police commission hearings" on page 59) if an allegation has been sustained.

Findings

The investigator writes a report presenting the results of the investigation and the factual basis for each recommended finding. After review by one of three senior OCC investigators, Mary Dunlap, the OCC director, reviews the file. Findings are based on a preponderance of the evidence. Although the city charter gives OCC the power to recommend specific discipline, it generally does not; OCC can influence the severity of the punishment by recommending that the police commission—with its authority to provide the most severe sanctions (see "Police commission hearings" on page 59)—hear a case.

Office administrative staff prepare and mail letters containing preliminary findings to each complainant and named officer. Either party may request an investigative hearing with an independent hearing officer granted at the discretion of the OCC director. OCC received 76 requests for investigative hearings in 1997, granted 12, and held 7 in 1997 and 5 in 1998.

Internal affairs division

If an allegation is sustained, OCC sends the case report containing a summary of the relevant evidence and law to internal affairs, whose staff decide whether they agree with the finding. Internal affairs agrees with OCC's findings about 90 percent of the time. When this occurs, an IA commanding officer determines the level of severity guided by the department's *Disciplinary Penalty & Referral Guidelines,* which recommends specific sanctions for specific types of misconduct. The IA officer sends the finding with the discipline recommendation to the

---

"ADDED ALLEGATIONS"

After explaining the complaint process and asking about the incident, the investigator also asks the citizen questions designed to determine whether the subject officer(s) engaged in misconduct that the complainant may not have identified or been aware of—for example, the investigator may ask, "Did the officer search your pockets or just do a pat search?" A juvenile might be asked, "When you were taken to the station, were you cuffed to a bench? For how long?" If the complainant is a woman, she might be asked, "Were you transported to the station in a van with men in it?" The information the complainant provides may form the basis for the investigator to charge the officer with "added allegations"—misconduct that is related to the complaint but that the complainant did not mention to OCC. (See chapter 6, "Resolving Potential Conflicts.")

Management Control Division commanding officer. If the commander agrees with the disposition, IA staff notify Prentice Sanders, the assistant chief, of the recommended finding and punishment. With rare exceptions, Sanders agrees with OCC's sustained findings and IA's disciplinary recommendation. Internal affairs writes the subject officer a letter offering the option of "a chief's hearing" or acceptance of the stated discipline.

When IA disagrees with an OCC finding, the IA officer-in-charge and the OCC investigator discuss their disagreement. About half the time, they reach a consensus. When they fail to reach a consensus, the OCC sustained disposition with which IA disagrees remains in the officer's file, but no discipline is imposed unless the OCC director asks the chief to submit the case to the police commission for trial. When the chief and OCC disagree on whether disciplinary action is appropriate in a case that OCC has sustained, the following procedure is used:

1. The chief returns the file to the OCC director explaining his disagreement.

2. The OCC director reopens the investigation, if necessary. If she determines on review that discipline is not warranted, the matter is closed. If she determines that discipline is warranted, she prepares and forwards a "verified" [i.e., by her] complaint to the chief.

3. If the chief files the verified complaint with the police commission, the commission may elect to hold a hearing on the disciplinary charges against the officer.

4. If the chief decides not to file the complaint with the police commission, he must tell the commission in writing. After reviewing the chief's and OCC director's decisions, the commission may order the chief to file the complaint. The commission decides whether to hold a hearing to try the charges in the complaint and, if the charges are sustained, to determine the discipline.

This procedure places the ultimate decision regarding disciplinary action in the hands of citizens if the police commission chooses to hear a disputed case.

Although there has been only one instance (still ongoing) in which the chief has been obliged to decide whether to file a verified complaint with the police commission, there have been instances in which the commission has ruled on disagreements that the chief and OCC have asked it to resolve.

> When officers in a hostage situation heard the hostage taker, who was holding a knife to the victim's throat, say he would kill her, they could have legally used lethal force. However, an officer threw his baton instead, hitting the man on the head and ending the crisis. OCC requested the officer be suspended because department orders prohibit throwing a baton. The internal affairs division said that using less-than-lethal force was preferable in the situation to using lethal force. Because IA and OCC disagreed on the finding, the police commission heard the case. The commission refused to sustain the allegation, ruling instead that the department's policies on use of batons needed to be changed.

The chief's hearing

Fred Lau, the current chief of police, delegates the chief's hearing to Prentice Sanders, the assistant chief. However, the chief reviews all of Sanders' decisions. Sanders upheld OCC's sustained findings and imposed discipline in 74 of 88 chief's hearings held in 1998.

The Management Control Division schedules and runs the chief's hearings with the subject officer, union representative, and the officer's captain present. An OCC attorney prosecutes the case. The union representative gives the subject officer's version of the incident and may introduce evidence exonerating the officer. The captain often gives an opinion about the case as well. Because of the informal nature of the chief's hearing, no sworn testimony is taken, although Sanders may ask the officer some questions. About half the time, Sanders makes a disciplinary decision at the hearing; the rest of the time, he decides later. Officers may appeal Sanders' decision to the police commission if a suspension of at least 1 day is ordered.

Police commission hearings

The police commission hears cases that:

- Subject officers have appealed and the commission agrees to accept.

- The chief forwards to it.

- Involve more than a 10-day suspension.

- The commission decides to hear because of a conflict between OCC and the chief on a finding.

- Involve driving under the influence and domestic violence (most of which are IA, not OCC, cases).

The commission holds a formal administrative hearing to redetermine the finding, this time at the hearing level (versus OCC's investigation level), and to impose punishment if the commissioners sustain the allegations. Commissioners first conduct a factfinding hearing and then receive and review transcripts of that hearing before a penalty hearing. At the final administrative hearing, all parties are present. An OCC trial attorney prosecutes the case, and the union attorney or privately retained counsel defends the officer. After opening statements, there is direct and cross examination of the parties and witnesses. In highly publicized cases, as many as 600 people have shown up to observe.

Commissioners, who deliberate in private, make their determination based on a preponderance of the evidence. The commission's findings often are unanimous. The commission can suspend officers for up to 90 days per offense or terminate them. Officers may request a judicial review to appeal the commission's decision. Of the 12 commission hearings held in 1998, 2 involved OCC cases. Commissioner hearings are relatively infrequent because, when officers agree to a suspension or resign rather than be fired, the hearing is canceled.

Other activities

OCC provides the police department with policy recommendations, arranges mediation, and assists with the department's early warning system.

Policy recommendations

OCC submits policy recommendations to IA and includes them in its annual report to the police commission. OCC's 1997 annual report provided 15 policy recommendations arising out of citizen complaints.

If IA agrees with an OCC policy recommendation, it tries to negotiate a solution with the OCC director—for example, restating an existing policy or requiring additional training. In 1997, OCC recommended that officers with complaint records be rejected as field training officers (FTOs). Internal affairs and the department compromised on a new policy that includes a review of the number of complaints against a candidate for FTO but excludes complaints from more than 5 years previous to the FTO's candidacy. The union and police commission both approved the change. Chapter 3, "Other Oversight Responsibilities," identifies additional policy recommendations that OCC has made.

Mediation

OCC provides a mediation option, but few citizens agree to the alternative, apparently because they feel uncomfortable with the approach. Of the 22 new cases eligible for mediation in 1997, 16 complainants (and 2 officers) refused to mediate. Twelve cases were mediated during the year (including several cases held over from the previous year). OCC uses volunteer mediators approved by the San Francisco Bar Association.

Early warning system

Every 3 to 6 months, OCC submits a report to the police department and to every commanding officer identifying the name and badge numbers of each officer who has three or more OCC complaints (excluding unfounded complaints) over the previous 6-month period or four or more complaints within the year. The report for the first half of 1998 identified 78 such officers. The first time their name appears, officers are given a performance review; the second time, getting a promotion and special assignments may be in jeopardy (and they cannot be a field training officer for 5 years).

Staffing and budget

In mid-1998, OCC had 15 full-time investigators (including two practicing trial attorneys) and a total staff of 30. Proposition G, approved by San Francisco voters in 1995, requires the city to hire one OCC investigator for every 150 police officers (see chapter 4, "Staffing"). As shown in exhibit 2–17, OCC's fiscal year 1998–99 budget was $2,198,778.

Distinctive features

The most unusual feature of San Francisco's oversight process is that an independent body in effect acts as the police department's internal affairs unit for citizen complaints about police misconduct.

EXHIBIT 2–17. OFFICE OF CITIZEN COMPLAINTS 1998–99 BUDGET

Budget Item	Funding Level
Permanent salaries–miscellaneous	$1,595,449
Overtime	10,323
Mandatory fringe benefits	373,339
Travel	1,500
Training	1,500
Membership fees	450
Professional and specialized services*	7,500
Rents and leases	119,500
Other current expenses	39,602
Materials and supplies	12,493
Other fixed charges	600
Services of other departments	36,522
Total	**$2,198,778**

* Contract fees to third parties, such as expert witnesses and translators.

- With the exception of complaints by one officer against another and incidents involving off-duty officers and nonsworn personnel, OCC alone conducts the San Francisco Police Department's investigations involving alleged officer misconduct. (The department and OCC investigate use of firearms simultaneously and independently.) This approach may increase the community's confidence in the independence of the oversight process. Some police feel that OCC investigators are not competent to evaluate their behavior.

- Organizations, not just aggrieved individuals, may—and frequently have—filed complaints with OCC. Allowing organizations to file expands the opportunity for the community to contribute to the oversight of police behavior. It may also encourage or enable groups with political agendas to try to influence the oversight process.

- Each OCC investigator's finding is reviewed by as many as three supervisors. Trained legal staff review every sustained case. Multiple reviews increase the opportunity for quality control. Multiple reviews also require extra time and expense.

- San Francisco voters approved an initiative that requires the city to hire one OCC investigator for every 150 police officers to ensure that there are adequate staff to address all citizen complaints. The required ratio also increases program costs.

- OCC's findings cannot be changed by the police department; only the police commission can overturn an OCC finding. The findings go into officers' files even if the department refuses to hand out any discipline.

- Citizens in San Francisco can make the ultimate decision on whether an officer is disciplined. If the chief and OCC director disagree on whether disciplinary action in a sustained case is appropriate, and the chief decides not to file the case with the police commission for a judgment, there is a process by which the police commission may elect to review the case and decide to hold a trial.

- In its investigatory capacity, OCC acts as a neutral party between the complainant and the police officer. However, if the case goes to a chief's or police commission hearing, OCC attorneys prosecute the officer. This dual role may blur the program's mission in fact or in the public's and police department's perception, resulting in antagonism from some community groups and the police.

For further information, contact:

Mary Dunlap
Director, Office of Citizen Complaints
480 Second Street, Suite 100
San Francisco, CA 94107
415–597–7711

Fred Lau
Chief of Police
San Francisco Police Department
Hall of Justice
850 Bryant Street
San Francisco, CA 94103
415–553–1551

Prentice Sanders
Assistant Chief of Police
San Francisco Police Department
Hall of Justice
850 Bryant Street, Room 525
San Francisco, CA 94103
415–553–9087

Tucson's Dual Oversight System: A Professional Auditor and a Citizen Review Board Collaborate

Background

In 1996, several Tucson police officers were sent to prison for assault, armed robbery, and child molestation. As a result, some citizen groups and complainants' attorneys felt the existing police oversight board, established in 1980, was not adequately supervising police misconduct. In response, the mayor asked the city council's public safety committee to present options to the city council for new forms of oversight. In March 1997, after intense debate, the mayor and city council replaced the old board with a new and more powerful Citizen Police Advisory Review Board. At the same time, they established a new position of independent police auditor. The council hired Liana Perez as the first auditor in July 1997.

From September 1, 1997, through June 30, 1998, 289 citizens contacted Perez. She or her staff answered questions from 155 of the individuals who called. The auditor's office took 96 formal complaints, which Perez forwarded to the police department for investigation. The remaining contacts were requests by citizens for Perez to monitor or review complaints they had filed directly with the police department. During this 10-month period, she monitored 63 ongoing investigations.

As explained in the following sections, the auditor and board have some overlapping responsibilities as well as different duties.

The independent police auditor

The city manager appoints the auditor to a 4-year renewable term. He can dismiss her, however, only with the approval of six of the seven city council members. The city manager meets with Perez every 2 weeks, and every month she submits a performance report to him. The auditor's office is located in city hall.

The auditor's principal responsibilities are to:

1. Serve as an alternative to the police department for accepting citizen complaints.

> **THUMBNAIL SKETCH: TUCSON**
>
> Model: citizens review cases (type 2) and audit IA procedures (type 4)
>
> Jurisdiction: Tucson, Arizona
>
> Population: 449,002
>
> Government: strong city manager, city council, weak mayor
>
> Appointment of chief: city manager
>
> Sworn officers: 865
>
> Oversight funding: $144,150 for the auditor, none for the board
>
> Paid oversight staff: two full time
>
> Tucson has both a full-time professional police auditor and a volunteer citizen review board. Both the Independent Police Auditor, appointed by the city manager, and a seven-member Citizen Police Advisory Review Board, appointed by the mayor and the city council, independently review completed IA investigations for thoroughness and fairness, and both make policy and procedure recommendations to the police department. The auditor also reviews cases when citizens appeal an IA finding, and she sits in on selected IA interviews to monitor the investigation process. The board acts as a pipeline for transmitting general community complaints to the police department. There has been no duplication of effort because the board typically asks the auditor to examine the cases it wants reviewed, and the auditor regularly attends and gives reports of her activities at board meetings.

2. Monitor ongoing investigations as needed by sitting in on internal affairs interviews.

3. Proactively audit—that is, review—completed IA investigations of citizen complaints for fairness and thoroughness.

4. Review cases in which a citizen expresses dissatisfaction with the police department's resolution of a complaint.

Intake

The auditor's full-time customer service representative accepts complaints in person, in writing, by facsimile, or by telephone. Citizen review board members (see "Citizen Police Advisory Review Board" on page 65), city council members, and community groups refer complainants to the auditor. The auditor forwards complaints to the police department's internal affairs bureau.

Monitoring

Exhibit 2–18 shows the auditing process and its relationship to the citizen review board. As shown, every week the police department forwards a list of new complaints to Liana Perez with the subject officers' names so she can decide whether to sit in on any IA interviews before the investigations have been completed. Perez monitors serious cases involving allegations of use of excessive force. She monitors other cases based on:

1. Random selection.

2. A citizen's request that she attend.

3. A request from the citizen review board to attend.

Sometimes Perez attends the interviews to make the complainant feel more comfortable—for example, if a female complainant wants another woman with her—or if a complainant feels he or she will be unable to articulate the complaint during a police interrogation. Perez sat in on one investigation when a woman came to her with a complaint that involved several officers. Because Perez knew that the officers' statements would be critical to a fair determination of responsibility, she wanted to ensure that each officer would be interviewed immediately after the other so they could not compare stories.

The city ordinance specifies that the auditor "cannot question witnesses but may suggest questions to be

EXHIBIT 2–18. CITIZEN OVERSIGHT PROCESS IN TUCSON

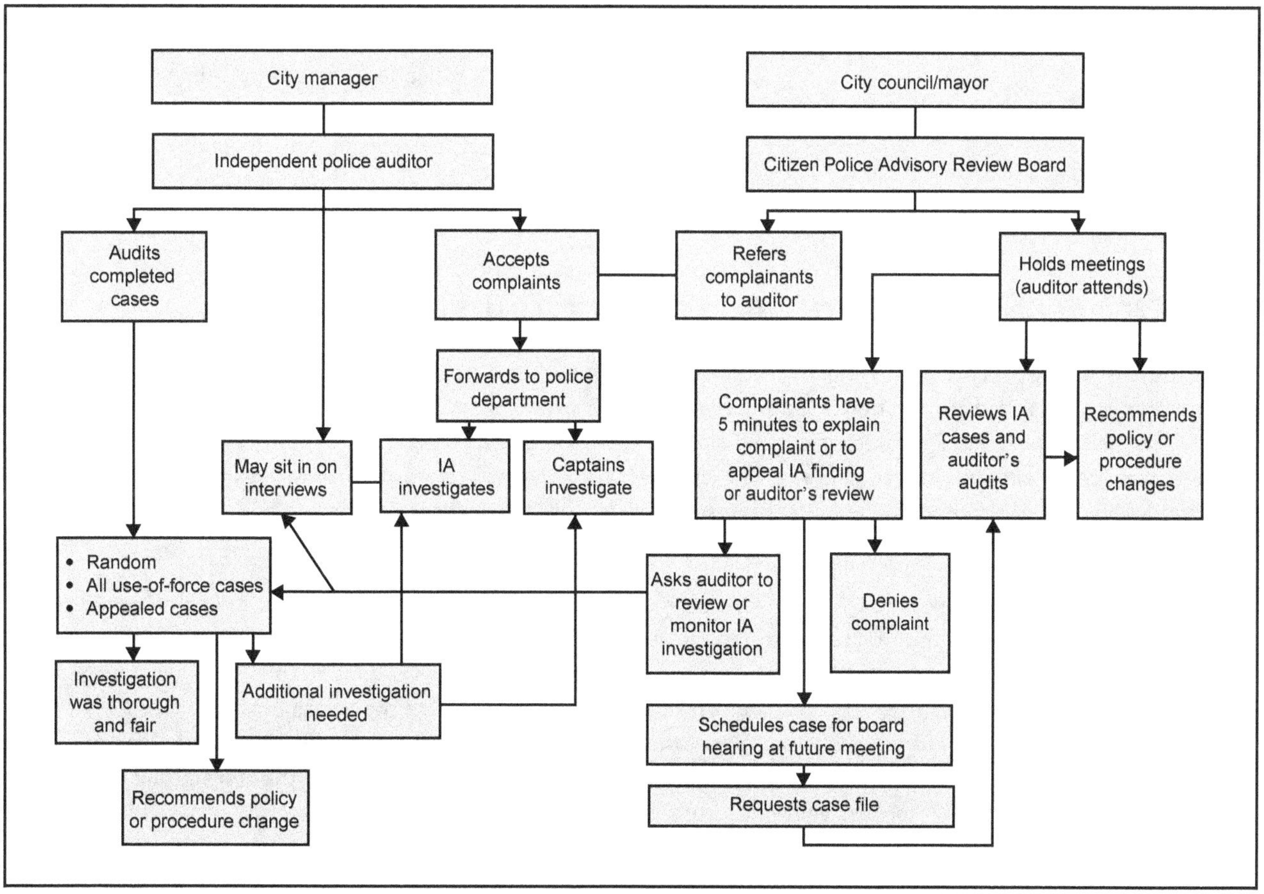

asked" by IA investigators (See "An Officer's View of the Auditor's Monitoring Activity"). Typically, Perez waits for a pause in the questioning and then says she has a question, such as, "I'd like him to clarify where he was standing when. . . ." The investigator then repeats the question to the subject officer or witness. Some IA investigators permit Perez to ask questions only at the end of each interrogation; others allow questions during the interview. Some investigators allow her to ask questions directly of the subject officer or witness.

The auditing process

Perez audits a random sample of completed investigations and all cases involving allegations of excessive force. The police department must comply if she requests additional investigation in a case. The auditor may also speak directly with civilian witnesses regarding the fairness and completeness of investigations.

The auditor does not sustain or disapprove of IA findings, and she does not make disciplinary recommendations. Instead, she reports on whether the investigation was fair and thorough. In effect, the auditor reviews IA's performance, not subject officers' behavior.

Perez has not had to talk with the chief or city manager to resolve a disagreement with IA. She did send back a case that IA did not sustain because she felt the investigating supervisor had disregarded key evidence implicating an officer. Internal affairs conducted additional interviews and sustained the allegations.

Other activities

Perez looks for patterns of complaints in her audits and telephones the IA commander if she finds a need for improvement. Perez expressed concern about supervisors repeatedly overlooking previous complaints against individual officers in deciding on discipline. As a result, the department formed, and invited her to participate on, a task force to examine how discipline was being meted out and how previous complaints should fit into the disciplinary decision. If the chief refuses to implement a policy recommendation, the auditor can appeal the refusal to the city manager, to whom both she and the chief report.

In fiscal year 1997–98, Perez arranged for four cases to be settled through informal mediation because the citizens did not want to file formal complaints but did want to express an objection to an officer's behavior. Perez brought the parties together and mediated the disagreements herself.

Staffing and budget

The auditor's office includes two full-time staff: Perez and an administrative assistant who takes citizens' initial complaints and has the authority to audit completed investigations.

AN OFFICER'S VIEW OF THE AUDITOR'S MONITORING ACTIVITY

The uncle of a suspect an officer had arrested filed a complaint with IA claiming the officer needlessly pointed a gun at him. The assigned IA investigator told the officer there would be an investigation and that Liana Perez, the auditor, would be present. The investigator told him Perez would have questions but the investigator would repeat the questions to the officer, who should then direct his answers back to the IA investigator, not to the auditor. Just before the interview, the union representative also told the officer to wait for the investigator to repeat each of the auditor's questions.

Perez asked several questions through the investigator to get a clear picture of what happened. The officer had been part of a team doing a high-risk stop of a homicide suspect who turned out to be the suspect's lookalike brother. Because five members of the department's gang intelligence unit had been involved in the incident, some of Perez' questions were designed to determine when each one arrived, where they were positioned, and what they did. Her goal was to determine whether any of the officers had seen the subject officer point the gun but had not admitted to it in their reports. Perez also asked questions to determine how far the subject officer was from the uncle, where the officer was going to secure the weapon (which belonged to the suspect), and whether there were any physical barriers to a clear view between the officer and the uncle. IA exonerated the officer.

Exhibit 2–19 shows the auditor's budget for fiscal year 1997–98—when the office was created—and for 1998–99. As shown, the startup budget was $144,150, with almost 68 percent allocated to staff salaries. In addition, $37,400 were allocated for what are likely to be one-time expenses, such as the purchase of office furniture and computer equipment and software. As a result, the requested 1998–99 budget is only $118,710, with 87 percent allocated to staff salaries.

Citizen Police Advisory Review Board

The Citizen Police Advisory Review Board has seven voting members. It also has seven nonvoting members—four community advocate members and one member each appointed by the city manager's office, police department, and police union. The board elected Suzanne Elefante as its first chairperson.

Board operations

As shown in exhibit 2–18, the city ordinance authorizes the board to:

1. Refer citizens who wish to file complaints to the auditor and the police department.

2. Request the auditor to monitor a particular case and present her findings.

3. Ask the police department to review a completed case.

4. Review completed IA investigations itself for fairness and thoroughness.

The board spent most of 1997–98 getting organized, including developing its procedures, establishing subcommittees, and training board members.

Intake

About eight citizens a month call their council representatives to complain about alleged police misconduct. Most council members give them the name and telephone number of the member of the review board whom they have appointed along with the auditor's telephone number. Board members, in turn, typically refer complainants to the auditor because the citizens are usually already unhappy with the IA investigation. When citizens file a complaint with the board, not with an individual member, Suzanne Elefante checks to see whether Liana Perez is already

EXHIBIT 2–19. TUCSON INDEPENDENT POLICE AUDITOR BUDGETS FOR FISCAL YEARS 1997–98 AND 1998–99

Expenditure	1997/1998 (adopted)	1998/1999 (requested)
Salaries	$77,220	$84,680
Fringe benefits	20,120	18,830
Public liability insurance	820	490
Office supplies	1,280	1,200
Hazardous waste insurance	50	60
Remodeling	20,000	500
Telephone	3,760	3,800
Duplication	3,500	3,500
Office furniture	5,000	1,000
Computers	7,000	0
Software	5,400	600
Maintenance of office equipment	0	180
Information technology	0	1,000
Memberships	0	350
Miscellaneous	–	1,020
Conference fees	–	1,500
Total	**$144,150**	**$118,710**

auditing the case. If she is, Elefante asks her to report her findings to the board; if not, Elefante asks her to audit it.

Appeals of completed investigations

Complainants who are dissatisfied with the IA investigation *or* the auditor's review may ask the board to review their cases. If the board agrees to review the complaint, it requests and receives IA's case files to examine between meetings. (State law makes IA investigations matters of public record.) The board may ask Liana Perez to do additional investigation or answer questions about the case if the auditor has already audited the case. After hearing from Perez, the board may recommend a different finding to the chief or the city council, but it has no power to enforce its recommendations.

Involvement of the public

The board meets from 6:30 p.m. to 9:00 p.m. the third Tuesday of the month in the main downtown library. The city clerk's office places notices of board and subcommittee meetings in newspapers and in city hall. The meeting begins with a "call to the audience" for any complaints and issues, with each person allowed to talk for up to 5 minutes. The board puts issues requiring more attention on the agenda for a future meeting, including complainants who wish to appeal an IA or auditor finding to the board. The auditor provides an update on the cases she has been receiving and monitoring. The board discusses other topics, such as the activities of off-duty officers. A police department IA member attends to answer questions about department policies and procedures and to report on IA activities during the previous month.

Other activities

The board may provide recommendations for changes in department policies and procedures to the chief, the auditor, the city manager, or the mayor and city council. Based on research by a board subcommittee that found that 11 of 26 police departments conducted random drug testing of all officers, the board sent a memo to the chief recommending random drug testing. The chief has not yet acted because testing is covered in the labor-management agreement, and the union is negotiating a new agreement.

Staffing and budget

Each council member appoints one board member for a 4-year term (or until the end of the elected official's term). The board reports to the mayor and the city council. The city clerk provides staff to type, duplicate, and disseminate the board's minutes. There are no other board expenses.

The relationship between the board and the auditor

The auditor and board do not officially report to each other. However, by city ordinance the board:

- Monitors the auditor by examining her monthly reports and asking her questions during monthly board meetings.

- May require the auditor to monitor or audit a case and report on her findings.

- May offer her recommendations.

The auditor and the board both have a legal mandate to review completed cases, either on their own initiative or in response to a citizen complaint. There has been no duplication of effort with reviews of IA investigations because the board typically asks Perez to audit cases about which it has a concern and to report back her findings; board members lack the time and expertise to conduct more than a few reviews themselves. There also has been no duplication because Perez has chosen to go to every board meeting, where board members can routinely ask for updates on her previous month's cases and the results of her monitoring activities.

Both the board and the auditor have made similar policy or procedure recommendations. For example, an officer who had had a personal relationship with a citizen took out a restraining order against the person and observed while another officer served it. Because there was no clear department policy prohibiting this specific action, IA exonerated the officer of the citizen's allegation of inappropriate behavior. However, after reviewing the case, both the auditor and the board independently requested the department to remind officers that they need to report to the department's legal department whenever they are involved in serving a restraining order in which the officer is a named party.

The auditor handles the day-to-day work of citizen oversight, while the board addresses general citizen concerns, not just complaints about specific acts of alleged misconduct. As a result, one board member feels "the board acts as the police department's eyes and ears for finding out the community's concerns about police behavior—it is the community's pipeline to the police." When citizens at a board meeting expressed concern that there was no random drug testing for regular police officers (except for narcotics officers), the board set up a subcommittee to research how other police departments conduct random testing (see "Other activities" on this page).

Distinctive features

The most unusual feature of the oversight procedure in Tucson is the use of both a paid, professional auditor and an independent volunteer citizen review board. (See "San Jose, California's, Independent Police Auditor Has Some Similarities and Differences With Tucson's Auditor.")

- According to José Ibarra, a city council member, "The dual system is good for constituents because it provides checks and balances." The board can act as a check on the auditor by the community to ensure that she is not operating as "just another government bureaucrat" rather than as a neutral but aggressive arbiter of complaints against the police. The auditor, in turn, provides the balance of ensuring that citizen complaints receive the

SAN JOSE, CALIFORNIA'S, INDEPENDENT POLICE AUDITOR HAS SOME SIMILARITIES AND DIFFERENCES WITH TUCSON'S AUDITOR

Thirty-six percent of all complainants in San Jose file their cases with an independent police auditor rather than with the police department. As in Tucson, Teresa Guerrero-Daley, the auditor, forwards the paperwork to the police department's internal affairs bureau, which conducts an investigation. The bureau then sends all its materials on all cases—including those filed directly with the police department—to the auditor. Exchange of information is simplified because the two agencies share a common computerized database.

Guerrero-Daley examines the case files for thoroughness and fairness, and she can request further investigation if she is not satisfied with a finding. She monitors selected cases by sitting in on interviews or going to the scene of the alleged incident. She becomes involved in all use-of-force cases. As can Liana Perez in Tucson, Guerrero-Daley can require the IA investigators to ask questions she may have of complainants and officers during interviews.

Command staff, not IA staff, determine a disposition after the investigation. Complainants who disagree with the finding or disposition may appeal to the auditor, who will review the case. If Guerrero-Daley disagrees with the disposition, she sends a memo to the chief. On the few occasions each year when she and the chief disagree, they meet together with the city manager (who appoints the chief) to reach a consensus. Guerrero-Daley can make specific recommendations for training as well as for changes in policy and duty manuals. The chief has adopted 90 percent of her recommendations.

With a staff of four professionals, the auditor's office has a budget of $320,000. There is no citizen review board in San Jose. Other cities with auditors—located primarily on the West Coast—include Seattle and Los Angeles. The Charlotte-Mecklenburg Police Department in North Carolina, on its own initiative, hired a private accounting firm to audit and recommend improvements to its complaints process.*

---

* See Walker, Sam, "New Directions in Citizen Oversight: The Auditor Approach to Handling Citizen Complaints," in *Problem-Oriented Policing: Crime-Specific Problems, Critical Issues and Making POP Work*, ed. Tara O'Connor Shelley and Anne C. Grant, Washington, D.C.: Police Executive Research Forum, 1998: 161–178.

---

concentrated and skilled attention that the board does not have the time or expertise to provide. The dual system also provides a check and balance in the sense that citizens can seek help from one office if they are dissatisfied with the other office's response. This may motivate each office to do an especially good job so that it is not second guessed by the other.

- The auditor and board complement each other in some respects:

  — The board provides for direct citizen involvement in police oversight, while the auditor represents city government. According to Capt. George Stoner, commander of the IA unit, "The dual system makes sure that the department addresses all segments of the city"—citizens and each branch of local government.

  — The board enables community representatives to offer the lay perspective of the citizen regarding IA investigations of alleged police misconduct, while the auditor's office makes it possible for a professional investigator to examine the department's investigations of alleged misconduct.

  — The board provides a public forum in which citizens can express general concerns about the department, while the auditor can address dissatisfaction citizens have about how IA handled specific complaints.

  — When the auditor and board agree on a recommended policy or procedure change, the recommendation in effect has the backing of the city manager, city council, and mayor. The auditor's and board's agreement on a recommendation also means that a professional investigator and lay citizens have agreed on the need for a change. This broad-based support may have more weight with the chief and local government than if only the auditor or the board propose the change.

— In time, a division of labor may evolve in which the auditor devotes most of her efforts to reviewing cases (as she already does) and the board focuses on developing policy recommendations.

- The dual oversight system provides citizens with two avenues outside the police department for registering dissatisfaction with IA investigations.

- The dual oversight system can involve some redundancy—that is, the auditor and board are both engaged in conducting audits and recommending policy changes, activities that either one of them could do effectively without the other. Redundancy could then impose an unnecessary financial cost on taxpayers. The potential for redundancy will be increased if one of the following occurs:

    — The auditor and chairperson of the board fail to cooperate because of personality differences, lack of interpersonal skills, an uncontrolled desire for publicity or power, or strongly held and antithetical views regarding the nature of police work.

    — Political dissension arises between the city manager, who appoints the auditor, and the city council, which appoints board.

- Neither the auditor nor the board has the legal authority to conduct investigations into alleged police misconduct themselves, and they do not have subpoena power. Rather, they audit, monitor, and publicize IA's conduct of investigations. By ensuring that IA investigations are done properly, the auditor approach may eliminate the need for independent professionals to investigate citizen complaints. This may reduce the cost of citizen oversight.

- Citizens have three choices about where to file—or refile—a complaint: with IA, the auditor, and the board. This approach enables complainants "to shop for the best deal." It also, in the words of a police officer, gives them "three bites at the apple": If they are dissatisfied with the finding by one office, they can take their complaint to the others.

The auditor's Web address is *www.ci.tucson.az.us/ia.html*. For further information, contact:

Liana Perez
Independent Police Auditor
Office of the Independent Police Auditor
255 West Alameda
First Floor, South
Tucson, AZ 85726–7210
520–791–5176

Suzanne Elefante
Chairperson, Citizens Police Advisory Review Board
2041 South Craycroft
Tucson, AZ 85711
520–790–4702

Capt. George Stoner
Tucson Police Department
270 South Stone Avenue
Tucson, AZ 85701
520–791–4441, ext. 1503

Teresa Guerrero-Daley
Independent Police Auditor, City of San Jose
4 North Second Street, Suite 650
San Jose, CA 95113
408–977–0652

Chapter 3: Other Oversight Responsibilities

KEY POINTS

- Citizen oversight bodies can undertake three other important responsibilities in addition to investigating, reviewing, or auditing complaints.

- Oversight bodies can recommend policy and procedure changes as well as training improvements.

    — Many experts regard this policy review function as the most important responsibility citizen oversight bodies can undertake because it can improve services throughout an entire department, not just among selected officers.

    — Many police administrators report that oversight bodies have made valuable policy and training recommendations that they have implemented.

- Oversight bodies can make mediation available to selected complainants. Minneapolis and Rochester make extensive use of formal mediation using trained mediators to conduct the sessions. Mediation can potentially benefit:

    — Complainants, many of whom are only interested in being able to express their concerns to the officer.

    — Subject officers, who can learn how their behavior can affect the public and can avoid having the complaint included in their files.

    — The community at large, as citizens improve their understanding of police operations.

    — Oversight bodies, which are spared the need to investigate and conduct hearings for these complaints. Mediation can have disadvantages and has limitations. For example, use-of-force cases are not suitable for mediation.

- Some oversight bodies assist police and sheriff's departments to set up or maintain an early warning system to keep track of complaints against officers who may need supervisory counseling or retraining.

In addition to investigating allegations of police misconduct and reviewing the quality of completed investigations, citizen oversight bodies can undertake three other responsibilities:

- Recommend policy and procedure changes and suggest training improvements.

- Arrange to mediate selected complaints.

- Set up or assist with the operation of an early warning system that identifies officers with potential problems.

Each of these activities has the potential to contribute to helping police and sheriff's departments remain or become accountable to the local community as well as to reduce police misconduct.

Policy Recommendations

"Many experts regard the policy review function as an extremely important aspect of citizen oversight. Policy review is designed to serve a *preventive* function by identifying problems and recommending corrective action that will improve policing and reduce citizen complaints in the future" (emphasis in the original).[1] Policy recommendations, including suggestions for training improvements, can influence an entire department, not just, as with oversight review, individual officers' behavior. According to Mary Dunlap, director of San Francisco's Office of Citizen Complaints, "Policy recommendations may be the most important work OCC does: They improve police services and the department's relations with the public."

Some police administrators believe that citizens do not have the necessary understanding of police practices to make useful policy recommendations. However, according to Capt. Gregory Winters, former officer-in-charge of the San Francisco Police Department's Risk Management Office (which includes the internal affairs unit), "The OCC's policy recommendations can be helpful precisely because they think of questions which, because the staff lack expertise [in police work], make you think." Adds Chief Fred Lau, "A lot of OCC's [policy] recommendations make sense, but the police department doesn't always realize they are needed."

> *Policy recommendations can influence an entire department, not just, as with oversight review, individual officers' behavior.*

The process of developing policy recommendations

Oversight bodies can identify the need for policy change in several ways.

- *Through individual citizen complaints.* The bracelet identification system described in "The Orange County Sheriff Implements a Citizen Review Board Suggestion" resulted from one family's complaint.

- *Through review of closed cases.* If the auditor in Tucson sees a need for a policy change, she can suggest the modification to the chief. The oversight board also can recommend the change to the chief. If the chief does not agree to implement the policy recommendation, the auditor can appeal to the city manager, to whom she and the chief report, and the city manager can require the change. The board can appeal the chief's refusal to the mayor and city council. As a result, when the auditor and board agree on a policy change, they have a great deal of potential clout behind them.

- *As a result of a general citizen concern.* Citizens in Berkeley may attend any regular Police Review Commission meeting or specially assembled public hearing to raise concerns that the board can use to develop a recommendation for a department policy change. One group of citizens petitioned for a public hearing to complain about the University of California campus police's use of pepper spray and batons to

THE ORANGE COUNTY SHERIFF IMPLEMENTS A CITIZEN REVIEW BOARD SUGGESTION

A mentally challenged man who had wandered away from his home during the night tried to enter a neighbor's residence thinking it was his own home. Officers who responded to the neighbor's call reporting a burglary in progress arrested the man, who spent 2 days in jail before his identity was discovered. The man's parents filed a complaint, but the board exonerated the two officers involved in arresting him, as did IA. However, the board recommended that the sheriff work with the county commission to develop a method to identify people with diminished mental capacity so they would not languish in jail for 48 hours.

Capt. Melvin Sears, the board's administrative coordinator, located a local mental health association that agreed to adapt its existing software to administer a program to distribute bracelets to these individuals. Kevin Beary, the sheriff, agreed to provide $1,600 from his forfeiture fund to purchase the bracelets, print an informational brochure, and purchase two Polaroid cameras to take photos for the bracelets. Later, Beary wrote the board, "It is always a pleasure to see positive results from an unfortunate incident. As you may recall, this is the result of the CRB case involving Mr. ____."

force demonstrating students out of the school's administration building, which they had occupied and refused to leave. As a result, the board developed recommendations on the department's use of pepper spray and batons that the city council endorsed and the campus police agreed to implement.

Examples of policy recommendations

Citizen oversight bodies can provide two general types of recommendations to change police operations:

- Changes in the way the department conducts its internal investigations into alleged misconduct.

- Changes in procedures that prescribe officer behavior.

Examples of both types of recommendations follow.

Recommendations for improving a department's own investigations of alleged police misconduct

Portland's oversight system has been especially active in recommending improvements to the police bureau's IA investigations.

- The Police Internal Investigations Auditing Committee recommended that IA handle all use-of-force complaints rather than send them to the precincts for investigation because of inconsistency of investigative quality at the precinct level. The department agreed.

- The auditor became concerned that supervisors were overlooking officers' patterns of complaints in deciding on discipline. As a result, the department has formed, and invited the auditor to participate on, a task force to address how discipline is being meted out and how patterns of complaints should fit into the disciplinary decision.

Lt. James Shepard, commander of Rochester's internal affairs unit, submits a form to board members to fill out and return assessing each investigation his sergeants conduct (see appendix A).

According to Capt. Gregory Winters, former officer-in-charge of the San Francisco Police Department's Risk Management Office, "The OCC's policy recommendations can be helpful precisely because they think of questions which, because the staff lack expertise [in police work], make you think."

Recommendations for improving policies governing officer behavior

Most of the oversight systems examined in this report have developed policy recommendations designed to improve officer conduct.

- In the wake of riots in a local park in 1991 and with more than 30 complaints from citizens regarding allegations of officer misconduct, the Berkeley City Council directed the Police Review Commission (PRC) to review and make recommendations on "all aspects of crowd control at large demonstrations." As a result of its study and deliberations, PRC recommended 12 specific changes that the department later implemented. The recommendations included:

  — Obtaining and using better amplified sound devices to address crowds and monitoring the audibility of dispersal orders.

  — Providing clearer instructions as to what specific location or area is the unlawful assembly site and the route by which persons will be allowed to leave and providing a reasonable opportunity for demonstrators to comply with the dispersal order.

  — Training specific officers to serve as crowd liaisons at demonstrations.

  — Barring the use of motorcycles as a means of force.

  — Proscribing the use of flashlights to harass or intimidate individuals in crowd control situations.[2]

- The Flint ombudsman's office saw that the department had no procedures for addressing victims after a domestic violence incident. The department agreed to develop a policy.

- The Orange County Citizen Review Board (CRB) agreed when IA did not sustain a complaint against a deputy who had used a "knee spike" to hit a noncompliant suspect in a specific portion of the body because the officer had special training in using this type of pain compliance technique. However, the board expressed concern that the agency's allowing a few trained deputies to use this technique posed a potential liability issue for the sheriff's office. The board observed that the department needed a clear written statement of when it was appropriate to use the kick and where the technique fell on the continuum of force. As a result, the sheriff's office developed a training bulletin that provides information on the relationship of the knee spike to the agency's use-of-force matrix. For example, the bulletin observes that "A knee spike can be used as a Level 3, 5, or 6 response. If the knee spike is used as a Level 3 response, the target area of the strike must be a large muscle mass, such as the outside portion of the thigh."

- The Orange County CRB expressed concern about deputies who return to duty after they have been involved in a shooting. As a result, the sheriff's office included new language in its use-of-force policy that states that the agency's staff psychologist will evaluate employees who have discharged their firearms before they are released back to full duty.

- St. Paul's review commission heard a case in which an officer failed to handcuff a suspect before putting him in the cruiser to transport him to the station for booking. The officer had difficulty extracting the person from the cruiser, and they got into a tussle. The officer had not followed the department's policy to handcuff every arrestee before transporting the person. However, officers and the board agreed that there are times when cuffing a noncompliant subject on the streets can excite the crowd. As a result, the department rewrote the wording of the handcuffing procedure to allow some officer discretion.

- In response to concerns raised by the auditor, the Portland Police Bureau chief issued the bulletin shown in exhibit 3–1 requiring officers to document in a report every use of handcuffs with individuals who are subsequently not arrested.

Mediation

A second important additional function some oversight bodies perform is to arrange for selected complainants to mediate their complaints with subject officers. In some cases, the mediation is informal. For example, the Tucson auditor occasionally tells an officer's captain or lieutenant that the complainant just wants to vent, especially when the officer did nothing wrong but simply did not explain his or her actions to the citizen. Typically, however, the process involves trained mediators who lead formal sessions at neutral locations.

Formal mediation: The process

Typically, the oversight body asks if the complainant is willing to mediate the complaint. If the person agrees, the oversight body directly or through internal affairs finds out if the officer is also amenable to mediation. If both parties agree, the organization that the oversight body uses to conduct mediations arranges a time for them to meet with a mediator in a private location. If the complainant expresses satisfaction with the result to the mediator at the end of the session, the case is considered closed. The content of the mediation remains confidential and typically nothing appears in the officer's file.

In some jurisdictions, complainants may not appeal the results of the mediation—that is, if they leave unsatisfied, they may no longer file a complaint with either the oversight body or the police department.

How two jurisdictions conduct mediation
Mediation is a major component of the Minneapolis and Rochester oversight processes. Although the content of the mediation sessions is similar in both cities, their oversight bodies arrange the process in a different manner.

Minneapolis. The Minneapolis Civilian Police Review Authority (CRA) refers appropriate cases to the Minneapolis Mediation Program. Pairs of trained volunteers mediate most of the sessions; program staff mediate the rest (see "Using Mediator Teams Has Advantages"). The Minneapolis Mediation Program requires volunteers to already be certified mediators and to attend its own 40-hour mediation course. New volunteers sit in on several sessions with experienced mediators before mediating sessions themselves.

If mediation is successful, the director dismisses the complaint; if it is not, she sends the case back to her staff for investigation. In Minneapolis, the parties reach agreement in about 90 percent of the cases.

EXHIBIT 3-1. PORTLAND POLICE BUREAU BULLETIN ON HANDCUFFING ISSUED IN RESPONSE TO AUDITOR'S RECOMMENDATION

CITY OF
PORTLAND, OREGON
BUREAU OF POLICE

VERA KATZ, MAYOR
Charles A. Moose, Chief of Police
1111 S.W. 2nd Avenue
Portland, Oregon 97204

September 4, 1996

TO: INTERNAL AFFAIRS DIV.
CAPT. JENSEN

FROM: 119/1526/CHO

READ AT ROLL CALLS AND POST

TO: All RU Managers

SUBJECT: Detention and Handcuffing Documentation

Recently we have received numerous citizen's complaints and tort claims notices regarding incidents where officers have temporarily detained and/or handcuffed individuals. When IAD or the City Attorney's Office have attempted to locate reports, none can be found. While we have generally been able to identify the officer through CAD files, the lack of a report causes us to have to begin an IAD investigation or a Risk investigation and interview officers. Many times we find that if a report would have been written, it would have contained enough information to decline the IAD investigation or to deny the tort claim with little or no inconvenience to the officer. Sometimes when we contact the officer regarding why they detained and/or handcuffed the individual, we find that the officer has the name of the individual in his/her notebook, but no details of the detention and it is difficult for the officer to remember the circumstances of the detention.

Therefore, effective immediately, whenever an officer: 1) uses any force to overcome resistance, to detain an individual, <u>or</u> 2) handcuffs and detains an individual, <u>and</u> 3) no arrest is made, the officer will document the detention/handcuffing in a report. While this may appear to be an inconvenience, I believe that in the long run it will mean less inconvenience. I have directed that the appropriate General Orders be revised to reflect this requirement.

Charles A. Moose
CHARLES A. MOOSE, Ph.D.
Chief of Police

c: DCA Woboril
DDA Pearson

DtnHndcf.CAM/ckf

USING MEDIATOR TEAMS HAS ADVANTAGES

Because two mediators facilitate every session in Minneapolis:

- There are always a man and woman mediator present.

- They can share perceptions about what is taking place and how to proceed.

- They can learn different mediation styles from each other.

- One mediator can pick up on verbal and behavioral cues the other may have missed.

- They can brainstorm on possible solutions when mediation reaches an impasse.

- They can debrief together afterward.

Exhibit 3–2 lists the program's mediation rules. As shown, the mediation proceedings are confidential except that the Minneapolis Mediation Program may inform CRA and the police department whether the parties met and reached agreement. Minnesota statute prohibits using mediation discussions and documents in subsequent legal or administrative proceedings.

Exhibit 3–3 is a copy of the form the parties sign. Two "terms of the agreement" that participants actually signed follow:

- "Both parties agreed that the dialogue was helpful in allowing them to understand each other's experiences and viewpoints."

- "The officer is sorry that the incident occurred and caused ____ embarrassment. . . . ____ acknowledges that the officer made the best decision possible with the information available on the scene."

Occasionally, a participant agrees to followup action:

> "Resources provided by [the complainant] will be forwarded to the Minneapolis Police training unit Room 204 City Hall for training/treatment of hypoglycemic diabetes with recommendation they be included in officer training."

One officer agreed to attend a cultural diversity course. Because there was no course available in the community, he attended the cultural diversity session the Minneapolis Mediation Program was running for its own volunteers. One complainant agreed to go on a ride-along.

Rochester. In 1984, a Rochester City Council member suggested that the Civilian Review Board (CRB) provide a conciliation option in an effort to help build positive relations between officers and citizens. Eight types of complaints are eligible for conciliation, such as "failure to take what complainant perceives was appropriate action." Cases involving allegations of use of excessive force are not eligible for conciliation.

CRB's parent agency, the Center for Dispute Settlement, has a pool of certified mediators it can tap for all its mediation components (e.g., victim-offender reconciliation). For mediating citizen complaints against the police, the agency chooses a mediator from among a subgroup who have participated in a 1-day extra training session on police conciliation.

The conciliation sessions are no different than traditional mediation sessions except that there is no written consent agreement between the parties at the end of a conciliation. At the end of the session, CRB sends a letter to internal affairs indicating, if the mediation was successful, that the case is closed and no investigation is needed. If the session was not successful, the letter informs IA whether the complainant still wishes to have the complaint investigated. In 1997, three of the four conciliations were successful. Of five conciliations conducted from January through September 1998, two were resolved, one was unresolved, and in two the complainant or officer did not appear.

Potential benefits of mediation

As summarized in exhibit 3–4, mediation can benefit everyone involved.

Potential benefits to citizens
- Mediation may encourage some citizens to file complaints who would otherwise be reluctant to come forward.

EXHIBIT 3–2. MINNEAPOLIS MEDIATION PROGRAM RULES

MINNEAPOLIS MEDIATION PROGRAM

Agreement to Mediate

The mediation process is voluntary. Any party may withdraw at any point if she/he feels that the mediation session is not serving her/his needs. Any oral or written agreement which proceeds from this mediation will generally result in compliance because the agreement represents those actions which all parties are willing and able to undertake.

You retain all legal rights to pursue this matter in other ways, but you cannot call a mediator to testify nor can you use mediation discussions, or meeting notes developed in or as a result of the mediation process, in subsequent judicial or administrative proceedings.

Rules of Mediation:

1. Each party will have an opportunity to present their side of the issue.

2. Each party is expected to mediate in "good faith."

3. During the mediation, any party or mediator may request to speak with another party(s) or mediator(s) privately (caucus).

4. Except for the mediated agreement or settlement, all communications and documents made during mediation proceedings are CONFIDENTIAL and cannot be published to third parties in writing or orally, except by the express written consent of both parties. The MMP may disclose to referring entities if the parties met, if an agreement was reached and if the agreement was complied with.

5. Minnesota Statute 595.02, Subdivision 1(k), makes testimony regarding any communications and documents made or used in the course of, or because of mediation inadmissible at subsequent legal or administrative proceedings.

6. A mediator cannot be called to testify at subsequent legal or administrative proceedings.

7. **Evidence of child abuse and/or abuse of a vulnerable adult will be reported. Also, threats of serious bodily injury directed at an individual or the substantial likelihood that an individual's actions, or inactions, may lead to the serious bodily harm of another, will be reported.**

We have read and understand the information presented here. We agree to work cooperatively to resolve any differences. We acknowledge that this agreement must be signed before the session can proceed.

Party	Date	Party	Date
Party	Date	Party	Date
Mediator	Date	Mediator	Date

2429 Nicollet Avenue South Minneapolis, MN 55404 (612) 871-0639

CHAPTER 3: OTHER OVERSIGHT RESPONSIBILITIES

EXHIBIT 3–3. MEDIATION SUMMARY AND AGREEMENT FORM

MINNEAPOLIS MEDIATION PROGRAM

SUMMARY AND AGREEMENT

THE FOLLOWING PARTIES: CASE # _____

    FIRST PARTY _____

    SECOND PARTY _____

    OTHER PARTIES _____

PARTICIPATED IN THE MEDIATION SESSION SUMMARIZED BELOW. THE SESSION TOOK PLACE

    ON _____ AT _____ AM/PM

    AT _____

THE NATURE OF THE DISAGREEMENT WAS:

_____

_____

_____

DURING THE COURSE OF THE MEDIATION, THE FOLLOWING TERMS OF AGREEMENT WERE REACHED.

_____

_____

_____

_____

_____

_____

_____

_____

_____

_____

_____

_____

SIGNATURE	DATE	SIGNATURE	DATE
SIGNATURE	DATE	SIGNATURE	DATE
SIGNATURE	DATE	SIGNATURE	DATE

THE SESSION WAS CONCLUDED AT _____ AM/PM.

THE PARTIES AGREE THAT ALL PROCEEDINGS WILL BE HELD IN STRICT CONFIDENCE.

White - Office   Pink - Participant   Yellow - Participant

EXHIBIT 3–4. POTENTIAL BENEFITS OF MEDIATION TO CITIZENS AND POLICE

Citizens may:
1. Be encouraged to file complaints.
2. Gain the satisfaction of talking directly with the officer.
3. Gain a better understanding of police work and why the officer acted in a specific manner.
4. Learn why some officers are not always courteous.
5. Feel more satisfaction than if a hearing results in an exonerated, unfounded, or not sustained finding.

Police officers may:
1. Learn how their words, behaviors, and attitudes can unwittingly affect the public.
2. Avoid having a complaint included in their files if mediation is successful.
3. Reduce the negative image some citizens have about officers.
4. Gain an understanding of why the complainant acted the way he or she did.

- Many citizens simply want the satisfaction of expressing their concerns face to face with the officer—letting the officer hear their side of the case or dissatisfaction with the officer's behavior. A survey of 371 citizens who had filed complaints with New York City's Citizen Complaints Review Board found that the desire for a direct encounter with the subject officer was "pervasive" and "significantly associated with complainant satisfaction."[3]

- Citizens can learn about the basis for police officers' actions in ways that can promote an improved understanding of the law enforcement officer's job.

  — Many mediated cases involve incidents in which an officer stopped and interrogated a suspect who turned out to be innocent, and the person became angry at having been "falsely accused" or singled out "for no good reason." Mediation lets officers describe how the information they had at the time led them to a reasonable suspicion that the person might have been the offender. (See "Two Successful Mediations.")

  — Officers often have considerable discretion in what they do, and citizens become upset when an officer chooses a course of action that is inconvenient (e.g., having their car towed) or embarrassing (e.g., pat searching them in front of neighbors). Alternatively, some citizens think officers have discretion in areas where they do not. For example, an officer may refuse to let a delivery truck drive down a street the driver normally uses to get to a retail store because the street is temporarily blocked off for a parade or local event. The officer could make an exception and let the driver go around the barricade, but, should the driver hit a pedestrian, the officer could be sued or disciplined.

- Like everyone, officers can be "having a bad day" and lose their temper with citizens. In addition, citizens do not realize how frustrating it can be when officers encounter repeated instances of citizen venality, venting at officers, or attempts to break the rules (e.g., driving in the breakdown lane). Officers are not justified in losing their temper and berating a citizen, but mediation can help citizens understand why officers did so.

- If the case is likely to result in an exonerated, unfounded, or not sustained finding by the review board, the complainant can feel better about a successful mediation than receiving one of these findings.

Potential benefits to officers

- Mediation can educate officers to the effects their words, behaviors, and attitudes can unwittingly have on the public.

- If mediation is successful, nothing negative appears in the officer's record. In San Francisco, any mediation the Office of Citizen Complaints schedules has this result even if the complainant fails to show up and as long as the officer appears. In Rochester, officers who agree to mediation do not have to go to IA to be interviewed or answer written questions. An IA commander in another city

Chapter 3: Other Oversight Responsibilities

TWO SUCCESSFUL MEDIATIONS

An officer was ticketing a car parked on the wrong side of the street when the owner came out of her house to complain. The officer ran the woman's name through the computer and found that a person matching her description had an outstanding warrant. The officer (a female) pat searched the woman and asked her to wait in the back of the cruiser. The officer then received more information indicating the woman was not the same person, so she released her.

The woman filed a complaint because she felt the officer had embarrassed her in front of her children. The officer, in turn, was angry she had to mediate the issue because she felt that, having done nothing wrong, the department should have told the woman the case was closed.

At the session, the mediator sat between them and asked them to decide who would talk first. The officer did, asking, "Was I rude?" "No." "Did I act professionally?" "Yes." The officer then explained why she had asked the woman to sit in the car, showing her the printout that indicated a person fitting her description—approximate age, race, gender, and same last name—had a warrant out for her arrest. The officer said, "I can understand why you were embarrassed, but if I was going to have you sit in the back of my cruiser, I needed to make sure you weren't carrying a gun that you could shoot me with in the back of the head." The woman became less frustrated and ended up satisfied with the officer's explanation.

* * *

The complainant reported he had been stopped for driving 45 miles per hour (mph) in a 30-mph zone but that the street was wide and deserted at the time. He said that the two officers in separate squad cars had yelled at him, pinned him against his car (so that his buckle scratched his new BMW), spread-eagled him, and did a pat-down search.

The officers present at the session explained to him, "You have to realize what we thought we were seeing. We had been chasing you with our lights and siren on for six blocks and you hadn't pulled over." The complainant responded that, precisely because the officers had made no other move to pull him over for six blocks, he was not sure they were signaling *him* to stop. The officer said he had not forced the man's car to the curb because there were pedestrians on the sidewalk for the first several blocks and the driver might have pulled onto the sidewalk and hit someone. "So I waited until we had reached some railroad tracks before forcing you to stop."

At one point, the mediators caucused so one could talk with the officer because he was starting to get angry. The mediator presented the citizen's point of view of feeling "violated" because he had no criminal record yet was being treated like a criminal. Upon their return, the officer explained, "We also thought that someone might have stolen the car, so we had to take precautions in case the driver was truly a bad guy." This explanation seemed to convince the citizen of the officer's good intentions. In turn, the officer could see that the man was only reacting to a frightening and inexplicable police action. The parties both apologized and signed a settlement.

tells all department personnel, "I can't tell you how to respond to an offer of mediation. But I can tell you that, if you go and it's successful, there will be no record of the complaint in your files. So what do you have to lose? You don't necessarily have to apologize or admit to wrongdoing, just explain why you did what you did."

- Mediation can help reduce the hostility and fear some citizens develop toward the police. Narcotics officers in Minneapolis raided an apartment looking for a drug dealer who, it turned out, was selling drugs only when the legal tenant and her three children were out of the building. Conducted at night with a no-knock entry with shotguns, the raid terrified the family. During mediation, the officers (who had done nothing wrong) apologized for the mistake and sat down and talked with the children so they would not be scarred by the experience to always be afraid of police officers.

- Just as mediation can give complainants an understanding of police behavior, some officers can benefit from learning the reasons citizens behave they way *they* do. For example, officers may learn that the only thing that upset the citizen was not immediately being given an explanation for why he or she was detained. As a result, mediation can enable some officers to learn what they can do differently that may reduce friction with the public.

Mediation may also benefit *the community at large*. According to Andrew Thomas, executive director of the organization that operates the Rochester Civilian Review Board:

Since IA and the Civilian Review Board both fail to sustain allegations in so many cases, it is important to move beyond assessing guilt or innocence in cases of alleged police misconduct to building a better understanding among citizens of what officers do and why they do it. If citizens gain an understanding about an individual officer's behavior, they may begin to understand all officers' behavior better.

Furthermore, if mediation results in improvements in officer conduct, the entire community benefits. Finally, mediation saves taxpayers the expense of an investigation and a hearing, or at least enables oversight and internal affairs staff to devote more time to more serious cases or reduce their backlog of cases.

MEDIATION IS NOT ALWAYS SUCCESSFUL

A woman filed a complaint with an oversight body because an off-duty officer checking ID cards at the door of a night club had confiscated her driver's license. He concluded it was fake because the woman could not identify the color of her eyes or the address listed on the card. Without the card, the woman was unable to pick up her disability check the next day. As a result, the woman filed a complaint. She wanted an apology and her license back.

After introductions, the complainant explained why she felt the officer had treated her disrespectfully by not believing the license was hers and, in general, "giving me a tough time" trying to enter the club. She said she wanted her license back. (The mediators felt she looked young enough to have possibly been underage.)

The officer talked for 3 minutes, saying that he was doing his job and was convinced that the ID was fake because the photo did not match the woman. He turned the ID in to the police department because it was standard procedure, and he no longer had the authority to retrieve it for her.

The mediators rephrased both their statements; for example, noting to the woman, "It's very important to you to get your license back." The woman explained that she had been caught up in the Department of Motor Vehicles bureaucracy trying to get it back, and began to cry. The mediators caucused, taking her in the hallway to give her a chance to calm down and telling her, "We understand that this was very upsetting for you."

Back in the meeting, the mediators asked the parties what they wanted to happen to have a satisfactory settlement. The woman repeated that the officer did not need to treat her the way he did and that she wanted her license back. The officer said again that he could do nothing about retrieving the license.

The mediators then caucused with the officer, who repeated that, because the ID was clearly fake, he would not apologize and he was justified in seizing it. Upon returning to the meeting, the woman became teary again and asked to end the mediation. Everyone stood up, the mediators thanked the parties for coming, and the officer and woman left.

Drawbacks to mediation

Mediation can have disadvantages (see "Mediation Is Not Always Successful").

- Because mediation is almost always held in private and the results are confidential, it may be seen as having less "teeth" than formal, public proceedings. For example, Robert Bailey, former Berkeley assistant city manager, believes that mediation circumvents the potential benefits of a public hearing that exposes officer misconduct to citizens and the media.

- According to Charles Moose, former chief of the Portland Police Bureau, some police administrators feel that mediation takes away their control over discipline because a condition of successful mediation is that there will be no further investigation and no discipline.

- Officers may go through the motions of appearing to be contrite to avoid having the complaint appear in their files. According to Todd Samolis, coordinator of the Rochester Civilian Review Board, "A number of officers come for the wrong reasons—to keep the complaint out of their file or to pacify their supervisors—but then they see it works: They hear the complainant in a new light and see how they might have handled the situation differently." Or they see that explaining their actions changes the complainant's attitude toward them. When Minneapolis Mediation Program staff telephone participants a day or two after the session, officers sometimes say, "I never really saw it from the citizen's perspective" or "My actions really *were* inappropriate; I was having a bad day."

- Although mediators are trained to make sure that each participant is on equal footing, some participants may have more of an advantage than others in certain situations.

    — Some officers are uncomfortable sitting in the same small room with someone whom they have cited or arrested.

    — Some mediators feel that officers appearing in uniform and armed may intimidate some citizens. However, some citizens report they prefer to talk to an officer who is in uniform so they feel they are not addressing an ordinary citizen but the person in his or her law enforcement role.

- In Tucson, complainants may opt for mediation, file a complaint with the auditor, file with the police department, or file with all three. Some police administrators object when mediation allows complainants to have more than "one bite of the apple" in this fashion. (However, in many jurisdictions, officers accused of misconduct also have multiple recourse, such as arbitration and civil service hearings.)

Mediation also has limitations.

- Mediation is suitable only for cases involving allegations of officer discourtesy and other minor misconduct. Allegations of use of excessive force or discrimination should not be mediated because, if sustained, they merit punishment.

> *Every 3 months, Minneapolis' Civilian Police Review Authority sends internal affairs the names of officers who have accumulated two or more complaints within the previous 12-month period.*

- Many officers refuse to participate in mediation. One officer said, "Why should I have to explain to a citizen why I did my job?" Some officers are reluctant to participate simply because, as an unknown procedure, mediation makes them nervous. According to Robert Duffy, chief of the Rochester Police Department, "Officers often find mediation threatening—people in authority have difficulty hearing the other side. But we need to hear why people disagree with us."

- Many complainants also are reluctant to participate. In 1997, San Francisco's Office of Citizen Complaints received 22 complaints that investigators felt were eligible for mediation, but complainants refused to participate in 16 of them.

Early Warning Systems

Early warning systems (EWSs) are procedures for keeping track of complaints against officers and using the results to target officers with unusually high numbers of complaints for supervisory counseling or retraining. The rationale for EWS was provided by a report that found a relatively small number of Los Angeles police officers were responsible for a disproportionate number of use-of-force reports and citizen complaints: Of about 1,800 officers against whom an allegation had been filed, 44 had 6 or more complaints against them.[4] Other studies have found that between 5 and 10 percent of a department's officers engage repeatedly in problem behavior.[5] As a result, in 1981 the U.S. Commission on Civil Rights recommended that police and sheriff's departments develop early warning systems to identify problem officers.[6]

Typically, EWS is designed to be informal, nonpunitive, and separate from the normal disciplinary process. Usually, it involves counseling or retraining by supervisory officers.[7]

Oversight involvement in EWS

Citizen oversight programs can become involved with an EWS in at least four ways:

- Recommend that the police or sheriff's department adopt an EWS.

- Collaborate with the department in implementing an EWS.

- Operate EWS for the department.

- Audit the department's EWS system.

After holding a hearing for a second complaint against a deputy for two separate shooting incidents, members of the Orange County Citizen Review Board learned that the deputy had a history of 18 disciplinary incidents. By making the lack of a tracking system public during its normal open hearing process and by stressing that the department was in jeopardy of lawsuits and negative media publicity by failing to discipline errant officers effectively, the board reinforced and sped up the department's existing plans to develop an early warning system. As of late 1998, the EWS software had been developed and was in place, and the sheriff's office was working on a policy to implement it.

An early warning system can help police and sheriff's departments identify officers who may be exhibiting a pattern of misconduct that suggests the need for intervention before the officers commit more serious misconduct.

Every 3 months, Minneapolis' Civilian Police Review Authority (CRA) sends internal affairs the names of officers who have accumulated two or more complaints within the previous 12-month period. A computer program generates the information. Internal affairs examines its own list of officers with multiple complaints and generates a report for commanders, the deputy chiefs, and the chief that identifies officers who have had two misconduct complaints of the same nature or three complaints of any nature combining both IA and CRA cases during the previous 12-month period. The report indicates what the complaints allege and whether they involve officers on duty or off duty. On average, about 12 officers per quarter have their names in the report. According to Lt. Dorothy Veldey-Jones, the IA commander:

Oftentimes a name will appear for one quarter and then never again. Occasionally, names appear for two quarters, but by the third quarter they drop off the list. Then, however, there are a few names that consistently appear on the list; they may drop off for short periods of time, but they seem to reappear frequently.

The department is evaluating the system to determine what the best courses of remedial action would be given the information the report provides.

The Portland auditor examined the Portland Police Bureau's EWS system and ensured it was identifying the individuals who met the bureau's criteria for inclusion on the list of potential problem officers.

Benefits and drawbacks of EWS

An early warning system can help police and sheriff's departments identify officers who may be exhibiting a pattern of misconduct that suggests the need for intervention before the officers commit more serious misconduct. However, departments must determine carefully how many complaints, what type of complaints, and the period of time that will trigger a specified supervisory action. For example, officers on drug details may have numerous complaints filed against them by drug dealers' attorneys in an attempt to intimidate the officers into less aggressive enforcement. Jurisdictions also must decide whether unsustained complaints will be included in the tally. While officers may object to this practice, one lieutenant reported that an officer who has accumulated 10 unsustained cases may indeed be getting into trouble, and, at a minimum, his or her supervisors need to be told to investigate whether there is a problem that requires corrective action before it escalates.

Police Accountability: Establishing an Early Warning System involves a national evaluation of early warning systems that discusses their benefits and limitations in detail.[8]

* * *

The success with which oversight systems are able to improve police and sheriff's departments' policies and procedures, conduct mediation, and assist with an early warning system depends crucially on the number, skills, impartiality, and dedication of their staff. The following chapter addresses the issues involved in staffing an oversight system.

Notes

1. Luna, Eileen, and Samuel Walker, "A Report on the Oversight Mechanisms of the Albuquerque Police Department," prepared for the Albuquerque City Council, 1997: 128–129.

2. San Francisco's Office of Community Complaints (OCC) also drafted a policy for crowd control in response to citizen complaints that officers were handling demonstrations by yelling "disperse" and then scattering the demonstrators by using their batons. The department adopted OCC's recommended policy that officers ensure that demonstrators have enough time to disperse and that there are enough avenues to leave the scene. Oversight bodies in both San Francisco and Berkeley may have been especially active in addressing their police departments' behavior in crowd control situations because of the unusual number of demonstrations the two cities experience.

Both cities also were working with the police to improve officers' handling of mentally ill persons. Officials in Albuquerque also are concerned about this problem. The National Institute of Justice, the research arm of the U.S. Department of Justice, has published an Issues and Practices report entitled *Police Response to Special Populations: Handling the Mentally Ill, Public Inebriate, and the Homeless* (Finn, P.E., and M. Sullivan, Washington, D.C., 1987, NCJ 107273) that describes efforts in 10 jurisdictions to enhance police and sheriff's departments' efforts to handle the mentally ill misdemeanor offender.

3. Vera Institute of Justice, *Processing Complaints against Police in New York City: The Complainant's Perspective,* New York: Vera Institute of Justice, 1989.

4. Independent Commission on the Los Angeles Police Department, *Report of the Independent Commission on the Los Angeles Police Department,* Los Angeles: City of Los Angeles, 1991.

5. "Kansas City Police Go After Their 'Bad Boys,'" *New York Times,* September 10, 1991; "Wave of Abuse Claims Laid to a Few Officers, *Boston Globe,* October 4, 1992, cited in Walker, Samuel, "Revitalizing the New York CCRB: A Proposal for Change," unpublished paper, Omaha: University of Nebraska, Department of Criminal Justice, September 1997: 5.

6. U.S. Commission on Civil Rights, *Who Is Guarding the Guardians? A Report on Police Practices,* Washington, D.C.: U.S. Commission on Civil Rights, 1981.

7. Luna and Walker, "A Report on the Oversight Mechanisms of the Albuquerque Police Department," 137.

8. Walker, Samuel, and Geoffrey P. Alpert, *Police Accountability: Establishing an Early Warning System,* IQ Service Report, vol. 32, no. 8, Washington, D.C.: International City/County Management Association, 2000.

Chapter 4: Staffing

KEY POINTS

- Citizen oversight bodies most commonly need three types of staff: volunteer board members, professional investigators, and an executive director.

- Talented and fair staff are essential for any oversight procedure to be effective.

- Because they may have no formal credentials, selecting board members is especially tricky.

  — Before recruiting board members, jurisdictions should establish the specific responsibilities they expect the board to assume. Then jurisdictions need to decide how large their board will be, members' terms of office, and their honoraria, if any.

  — A common selection criterion is to include diversity. Permitting current or former police officers or sheriff's deputies to serve is controversial.

  — The process of selecting board members can involve public hearings, private interviews, and word of mouth.

  — Training for board members can include lectures, materials review (e.g., department policies and procedures), attending a citizens' academy, ride-alongs, and training as mediators.

- Some oversight systems involve the use of paid investigators.

  — Investigators need to be able to handle the potential stress of interviewing sometimes angry complainants and hostile officers.

  — Many jurisdictions try to hire investigators with a law enforcement background.

  — Senior staff train new investigators. Novices also learn on the job.

- Along with the police chief or sheriff, the executive director or auditor will have the greatest influence on whether the oversight system achieves its objectives. Most jurisdictions make considerable use of word of mouth to find the most qualified individual.

Citizen oversight procedures may require three principal types of staff:

- Volunteer board members.

- Professional investigators.

- An executive director.[1]

Talented and fair staff in all of these categories are essential for citizen oversight to achieve its potential benefits. Incompetent staff will not be able to perform their responsibilities, while biased staff will create conflict that can grind the process to a halt. Staff must also be flexible. Fred Lau, the San Francisco police chief, said:

> The OCC [Office of Citizen Complaints] executive director is extremely critical to the relationship to the police department. If I didn't have a good working relationship with Mary Dunlap, it would be horrible. She is willing not to be rigid, sit down with department subject matter experts, and talk. She is tough and tenacious, but we have never come to an impasse—we've never had to go to the [police] commission for a tiebreaker.

As a result, the recruitment, selection, and training of oversight volunteers and staff are extremely important. The following discussion examines these processes for each of the three staff categories. The issue of staff supervision is discussed in chapter 7, "Monitoring, Evaluation, and Funding."

Volunteer Board Members

If a jurisdiction chooses to have volunteer citizens review cases, the volunteers need to be chosen with particular care because they usually have no formal credentials in the law enforcement field, may have inappropriate motives for serving, and may be viewed as especially unqualified by some police and citizens.

Planning decisions

There are several early decisions jurisdictions must make before recruiting volunteer board members.

- What will be the board's specific responsibilities—that is, what will members be expected to accomplish? The nature of their assignments and how much time they will need to achieve them in part will influence how jurisdictions answer the other planning questions listed below. For example, if board members will be expected to review—and, especially, hold hearings—on less serious cases of misconduct (e.g., verbal abuse) rather than focus exclusively on serious cases (e.g., use-of-excessive-force allegations), program planners need to provide adequate staff to avoid long delays in case processing.

- How many members will the board have? There appears to be no correlation between board size and the population of the communities they serve.[2] Most boards have between seven and nine members. Factors to consider in deciding on the number include:

  — Not having so few members that, with two or three absences, there is no quorum.

  — Having enough members to represent the diversity of the local community.

  — If small groups of board members will be conducting hearings (as in Berkeley, Minneapolis, and Rochester), having enough members so that the burden of holding hearings is not overwhelming.

The St. Paul mayor, with the consent of the city council, appoints three alternative commissioners to serve in the event a regular member does not attend a meeting or hearing due to illness or a conflict of interest.

- What will be the members' term of office, and can they be reappointed? Terms should not be so short that board members leave as soon as they gain valuable on-the-job experience[3] but not so long that undesirable members can remain in office beyond their welcome. Portland appoints members for 2 years, with the option of reappointment. Board members in Minneapolis and Tucson serve for 4 years. In Berkeley, new council members may replace previous members' selections before their 2-year term is over. In Tucson, board members may *not* serve beyond the term of the mayor or council member who appointed them.

- Will board members receive an honorarium? If so, how much? Some compensation may be necessary to attract qualified volunteers as well as to underscore that the community feels their work is important. Too much compensation not only becomes expensive but may also erode the concept that independent citizens, rather than paid staff employed by the city or county, are overseeing police misconduct.

  — Board members in Minneapolis receive $50 for each day they attend one or more meetings or hearings or provide other board-related services.

  — Panelists in Rochester who chair review meetings receive $50 for each 2-hour block of time they participate (they receive another $50 if they run at least 15 minutes into the next 2-hour block); regular panelists receive $35. Mediators receive $35 for each case. (See "Rochester's Board Members All Are Trained Mediators.")

  — Board members in Berkeley have been given $3 an hour (not to exceed $200 per month) since the

Talented and fair staff are essential for citizen oversight to achieve its potential benefits.

Police Review Commission was established in 1974. Board members in Omaha receive no compensation.

Selection of board members

Jurisdictions need to recruit and screen board members carefully.

Oversight legislation in Orange County, as in many other jurisdictions, requires that "The composition of the CRB [Citizen Review Board] shall endeavor to reflect the ethnic, racial and economic diversity of Orange County."

- The five civilian members of the St. Paul Police Civilian Internal Affairs Review Commission include one woman, one African-American, one Hispanic, and one gay person. They include the director of a community-based organization, the vice president of a lighting fixture company, a court psychologist, the director of enforcement for the State Commerce Department, and an IBM project director.

- The seven Minneapolis board members consist of three African-Americans, one Native American, and three Caucasians. Three members are women. Members' occupations are the former assistant State ombudsman for corrections, a minister, a professor who teaches police ethics, a retired social worker and probation officer, a school teacher who is also a nonsworn Parks Department police agent, a former city housing authority employee, and a county public defender's office employee.

As Minneapolis' board composition suggests, many jurisdictions look for volunteers with some type of background in the criminal justice system. Portland's citizen advisers include a retired State patrol officer, a retired police chief, a judge, and a defense attorney. Allowing current or former law enforcement officers to serve is controversial (see "Should Police Officers and Sheriff's Deputies Serve on Boards?"). Few jurisdictions choose active police officers to serve; some local ordinances prohibit their selection. The Berkeley ordinance prohibits all city employees from serving.

Allowing current or former law enforcement officers to serve on volunteer oversight boards is controversial.

Boards typically do not include local activists, such as members of local chapters of the American Civil Liberties Union or the Lawyers Guild.

Lisa Botsko's written description of Portland's citizen advisers' responsibilities and duties includes the ability to:

- Work with persons of opposing viewpoints.
- Provide constructive criticism.
- Communicate effectively, verbally and in writing.

ROCHESTER'S BOARD MEMBERS ALL ARE TRAINED MEDIATORS

Rochester trains all Civilian Review Board members in mediation. According to Todd Samolis, the CRB coordinator:

Mediation training exercises focus on helping participants to become aware of their biases—since everyone has them and they cannot be eliminated—so that as board members they can keep these prejudices in check when they review IA cases. Mediation training increases their ability to think impartially.

According to one board member, "The [mediation training] program was probably the most educationally enlightening experience I've ever had."

Samolis also believes that "mediation training increases listening skills dramatically. In addition, it helps panelists to absorb the information in the case files in terms of who said what, when, and where—to keep things straight—and to spot inconsistencies."

CHAPTER 4: STAFFING

SHOULD POLICE OFFICERS AND SHERIFF'S DEPUTIES SERVE ON BOARDS?

The inclusion of law enforcement officers (whether current or former) on volunteer oversight boards is a controversial issue. On the one hand, the St. Paul oversight ordinance requires the city's board to include two active police officers. (The St. Paul police union lobbied to have one officer for each citizen board member but settled for two officers on a seven-person commission.) Omaha's Citizens Complaint Review Board has one police union member. On the other hand, Tucson's ordinance prohibits board members from being current peace officers, while Berkeley's forbids current or former officers from serving.

Many people argue that having one or two officers on the board provides additional insight into police behavior. According to a former board member in St. Paul, "It is good to have officers on the board because they have a perspective citizen commissioners don't have; you want their frame of reference. But [they cannot exercise undue influence because], with only two members, they do not have a majority vote."

The Rochester board had two, then one, and now no officers as board members. One long-time citizen member reported: "I liked having an officer on the panel because, regardless of how much training civilians get, the officer is better versed in department policies. And they took their jobs very seriously." On a few occasions, police members drew other board members' attention to improper procedures that subject officers had engaged in which were not among the reasons for the citizens' complaints.

Opponents of allowing active or even former officers to serve as board members argue that their participation violates the concept of independent review: Officers, even if from other departments or retired, may not be able to be objective about the culpability of another officer's conduct. The Orange County charter establishing the Citizen Review Board is silent on the matter of whether the sheriff's two appointees to the board may be deputies. However, the sheriff has deliberately selected civilians to avoid any impression that the board is biased in favor of the department. Indeed, to make sure the board remained neutral in the public's perception, when a county commissioner nominated a correctional deputy to the board, the sheriff successfully asked the commissioner to withdraw the nomination.

A board member in Minneapolis added:

> All you need [to be a competent board member] is to be a citizen of sound judgment.... You don't need to understand police work to know if someone is mistreating someone else, such as calling them names. Abuse is obvious. And the hearing brings out whether the officer violated department policies or procedures.

A police officer observed, "I have heard comments from street cops that board members are clueless, but then I hear the same thing said about IA investigators who haven't been on the streets for years."

- Maintain confidentiality of information in IA files.
- Realize possible conflicts of interest such as relatives working with law enforcement.

Recruitment

Most jurisdictions recruit board members through public announcements and by word of mouth.

- Minneapolis has an open appointment process in which the city council's Public Safety and Regulatory Services Subcommittee hosts public hearings at which applicants present themselves. The subcommittee makes recommendations to the full council for approval by majority vote.
- In Orange County, citizens can tell county commissioners they would like to serve on the oversight board.

In addition, because there are so many advisory boards in Orange County, there is a board whose only task is to look for people to serve on other county boards.

Some jurisdictions experience difficulty recruiting board members because of the time commitment involved to be trained and to serve. Todd Samolis, coordinator of the Rochester board, had to court candidates because of the requirement to attend 48 hours of police academy training in the middle of August. A board member in Berkeley reported she spends about 50 hours a month reading materials and attending hearings. Board members in Orange County devote an average of 10–16 hours a month, and in Minneapolis the average is 10 hours a month. Most board members elsewhere spend a minimum of 4 hours a month.

Training

Training requirements for board members differ. City ordinances in Tucson and St. Paul specify that board members must attend mandatory comprehensive training before they may review any cases. The Tucson ordinance identifies nine areas of required training, from police department operations to confidentiality. The police department and independent auditor provide the 40-hour training. Some of the more common training methods follow.

Lectures
Barbara Attard, Berkeley's Police Review Commission (PRC) officer, runs a 4-hour session on PRC procedures that includes presentations by the chief on the discipline process and by the city attorney on open meeting regulations. Melvin Sears, the Orange County Sheriff's administrative coordinator, trains new members by reviewing the department manual and board manual, paying special attention to use-of-force issues.

Materials review
Board members typically are provided with written materials that include department general orders and other policies and procedures. Melvin Sears gives all new board members in Orange County a large notebook that details their responsibilities and includes many of the department's general orders.

> *Candidates for Rochester's board must attend a 2-week condensed version of a police academy.*

Citizens' academy
Candidates for Rochester's board must attend a 2-week condensed version of a police academy. Run by the police department, the 48-hour course involves 3 hours per evening for 2 weeks and two all-day Saturday sessions. The training includes using sidearms with a "Shoot/Don't Shoot" simulator, practicing handcuffing, and learning about department policies and procedures, such as the use-of-force continuum. According to one board member, "I had never fired a gun before. At first it was a strange sensation. But it helped me understand how inaccurate handguns are and the officers' need for split-second decisionmaking."

By ordinance, new board members in St. Paul may not be sworn in until they have completed the 11-week, 33-hour citizens' academy that includes getting sprayed with a minor dose of pepper spray, using a baton, handcuffing each other, and firing handguns using a "Shoot/Don't Shoot" simulator. The Albuquerque city ordinance requires commission members to attend the citizens' police academy. Most Orange County board members have attended a 36-hour citizens' academy on their own.

Using a citizens' academy as a training tool is not without controversy. The Minneapolis Civilian Police Review Authority was originally reluctant to have its board members attend the citizens' academy because of concerns that they might be "coopted" as a result of the process. By contrast, the police union wanted attendance to be a requirement for board membership. As of 1998, six of the seven board members had voluntarily attended the 12-week course. When Paul McQuilken, chairperson of the Orange County board, recommended that the county commission select board members only from among individuals who had already attended the academy, some community groups objected because they felt members would become too sympathetic to the police. A board member in another jurisdiction downplayed this risk: "We don't get brainwashed to believe blue."

The real issue, according to Mark Gissiner, president of the International Association for Civilian Oversight of

Law Enforcement (IACOLE) from 1995–99, is whether the empathy for police work that board members may develop by attending citizens' academies (not a bad thing in itself) makes it difficult for them to focus in the future on whether officers violated department policies and procedures.

Ride-alongs

In Minneapolis, board members must do one ride-along after they have been appointed. However, a police lieutenant in another city observed, "One or two ride-alongs are useless [as a learning tool] because officers are on their best behavior," and any one shift could be atypical. "Volunteer board members need to go on several to begin to gain an understanding of police work," he said. Reflecting this judgment, the St. Paul ordinance requires new board members to go on at least 10 ride-alongs, 2 in each district and 1 with each of 4 specialty units (e.g., traffic, search warrants, canine).

Evelyn Scott, a board member in Rochester, went on a ride-along and reported:

> I ended up running through back alleys and backyards following an officer chasing a suspect. The officer arrested the person, handcuffed him, put him in the cruiser, and drove him to the booking area. When the suspect kept cursing the officer the whole way, I realized how much patience officers have to have. Later [when she was a board member], we had a case in which an officer stopped a suspicious person. When the officer tried to frisk him, the man took off. The officer chased him, and the man fired back. The officer then shot and wounded the man. The citizen filed a complaint against the officer for use of excessive force. Reviewing the case, I remembered my ride-along and recalled how fast things happen, how quickly officers have to react, how situations that look routine may be dangerous, and how officers may have to make an instantaneous decision about whether to shoot.

> *The St. Paul ordinance requires new board members to go on at least 10 ride-alongs, 2 in each district.*

On-the-job training

In some jurisdictions, the bulk of the training occurs on the job; in all jurisdictions, some of the required experience can be learned only by doing it.

Inservice training

The Albuquerque city ordinance requires board members to attend a yearly 4-hour training session conducted by a civil rights attorney. Tucson legislation requires board members to pursue 48 hours of educational opportunities annually, such as ride-alongs and the police department's citizens' academy. The Orange County Sheriff's Department periodically provides members with an hour of inservice training before (or instead of) regular board meetings that has included explanations of:

- Deputies' procedures for dealing with armed and unarmed subjects in relationship to body shape and size.

- The procedures the department's psychologist follows in conducting fitness-for-duty evaluations.

- IA operations and chain-of-command procedures for reviewing investigations.

- Office policy and procedures related to its use-of-force matrix and defensive tactics that included simulated demonstrations by deputies followed by board member participation in exercises designed to help them determine the level of force used.

Investigators

For oversight systems that investigate alleged officer misconduct, selecting and training investigators also requires careful attention. According to Mark Gissiner, former IACOLE president and a senior human resources analyst for Cincinnati who has investigated allegations of police misconduct since 1985: "The investigation, analysis, and determination of whether excessive force occurred is an extremely difficult task."[4]

Recruitment

Mary Dunlap, director of San Francisco's Office of Citizen Complaints, requires candidates for investigator positions to have 2 years of investigative experience, which may be in academic research. Dunlap also looks for individuals who can handle the stress of angry complainants and hostile, armed officers. She tests applicants for tendencies to jump to conclusions, and she interviews them to detect biases for or against law enforcement. (See "San Francisco Mandates the Number of Oversight Investigators.")

Investigators in many jurisdictions have a law enforcement background. In Minneapolis, two of three current investigators are former police officers with other departments. According to Robin Lolar, one of the investigators:

> My police background enables me to detect when officers aren't being truthful in their reports by way of their creative writing. I can also sense when complainants are leaving something out of their stories. I know what officers can and can't do by way of stops and seizures. Knowing proper police procedure saves me a lot of research time.

Lolar added, "Complainants feel comfortable knowing I am a former police officer from outside of Minneapolis." (Minneapolis legislation forbids any present or former city officer from becoming a Civilian Police Review Authority [CRA] investigator.) The investigators' previous experience as police officers helps address police union and subject officers' concerns that CRA does not understand police work or is biased against officers. In fact, an IA sergeant who was exonerated of misconduct by CRA reported, "The investigator questioned me for 45 minutes and was very thorough and fair—in fact, I ended up hiring him as a criminal investigator for the city attorney's office."

An auditor's report on the Kansas City, Missouri, Police Department questioned how independent the city's Office of Citizen Complaints was from the department because three of the five staff members had ties to the department; two were former police officers, including one who was a former department IA investigator. The police commission resolved not to hire former officers again.

Oversight bodies must also consider carefully whether to hire investigators who are members of activist groups. Even if activists are able to be objective on the job, their volunteer activities off the job may

> *In 1996, San Francisco voters approved Proposition G, which amended the city and county charters to require that the Office of Citizen Complaints have at least 1 investigator for every 150 sworn officers.*

SAN FRANCISCO MANDATES THE NUMBER OF OVERSIGHT INVESTIGATORS

Inadequate funds to provide for sufficient staff can doom an oversight system because either investigations cannot be conducted thoroughly or cases will be delayed—or both. As discussed in chapter 5, delays result in disillusioned complainants and angry police officers as well as loss of memory and witnesses.

To avoid these shortcomings, in 1996 San Francisco voters approved Proposition G, which amended the city and county charters to require that the Office of Citizen Complaints (OCC) have at least 1 investigator for every 150 sworn officers. As a result, in 1998 OCC had 15 investigators and 4 supervisory personnel to handle a department with 2,100 sworn officers. By contrast, Minneapolis' Civilian Police Review Authority has only 4 investigators for a police department with 919 sworn officers.

However, even with the charter amendment, it took San Francisco many months to provide the money for OCC to hire the required investigators. And even with the increased staffing level, OCC staff continue to be overworked—each has 40–60 cases at any one time. Because officers generally may be interviewed only while they are on duty, investigators frequently conduct interviews at 6:00 a.m. if officers are working night shifts. Investigators are granted compensatory time for working before or after hours, but not overtime pay.

create the perception—as might the use of current or former police officers—that investigations may be biased.

Training

In Minneapolis, the director trains new investigators, who then sit in on cases handled by the senior case investigator. Investigators have attended inservice training conducted by police officers and others in use of force, verbal judo, search warrants, cultural diversity, and domestic abuse; they also have participated in seminars with a professional training firm on investigation, interviewing, and interrogation techniques. In San Francisco, the Office of Citizen Complaints' (OCC's) director, chief, and senior investigators, using a standardized training manual OCC managers developed, lead 8 to 10 full-day training sessions, followed by several weeks of working side by side with supervisors who monitor and correct their intake interviews, complaint analyses, witness searches, and officer interviews. The office follows up with two to four trainings each month for all staff on a variety of subjects.

Lisa Botsko, Portland's first police auditor, developed a set of "Standards of Review" that advisory board members are instructed to follow in conducting reviews of IA cases and determining whether the oversight body needs to review a case. The standards include guidelines related to the filing and intake process for complaints, investigations, and findings. (See appendix B.)

Although she is not an investigator, San Francisco's OCC policy and outreach specialist regularly attended recruit classes for 28 weeks at the police academy to improve OCC's knowledge of police department basic training and to establish rapport between OCC and recruits. "The bank of knowledge built by attending the academy," she said, "is vital to understanding police procedures."

Executive Director or Auditor

The executive director (or auditor), along with the police chief or sheriff, is the single most important person for ensuring that the oversight process is effective. Hiring or appointing experienced individuals is critical to establishing or maintaining the system's credibility. For example, Lisa Botsko, the Police Internal Investigations Auditing Committee's auditor in Portland from 1993 to 1999, had been a private investigator for insurance fraud companies and had conducted high security clearance investigations for the Federal Government's Office of Personnel Management in its Denver regional office.

Most jurisdictions send out public notices when they are hiring an executive director, but they also rely heavily on word of mouth to help identify the most qualified individuals.

- The Berkeley city manager hired Barbara Attard because of her reputation as an effective senior investigator for many years with San Francisco's Office of Citizen Complaints.

- In Minneapolis, the Civilian Police Review Authority (CRA) president hired Patricia Hughes, the current CRA executive director. The CRA chairperson, Daryl Lynn, had previously hired Hughes as a counselor in 1975 to work in a pretrial diversion program. Later, Hughes became an attorney and Lynn moved to another position. Serendipitously, Lynn became a paralegal with the Minnesota State public defender's office at a time when Hughes was an attorney in the office, so he was able to see her litigation skills firsthand.

Word of mouth can be the best method of hiring staff because jurisdictions more easily can identify individuals who are likely to be appropriate for the position than if they have to rely exclusively on resumes and interviews. Echoing what Police Chief Fred Lau said in San Francisco, Capt. Melvin Sears, the Orange County sheriff's board administrative coordinator, confirmed, "Who the people are is critical to the system's working."

Notes

1. In larger jurisdictions and in systems with a large volume of cases, programs also will need administrative and clerical support staff as well as data entry personnel. If the system prosecutes cases, it will need attorneys. San Francisco's Office of Citizen Complaints employs a policy and outreach specialist.

2. Walker, Samuel, *Citizen Review Resource Manual*, Washington, D.C.: Police Executive Research Forum, 1995: 11.

3. A police officer said, "Board members don't understand police work—how volatile and ugly bad guys are and the need to act quickly to avoid escalation. But once on the board a while, they develop a sense of what takes place on the street."

4. "Use of Force," paper presented at the 1995 International Association for Civilian Oversight of Law Enforcement World Conference, Vancouver, British Columbia, Canada, 1995.

Chapter 5: Addressing Important Issues in Citizen Oversight

KEY POINTS

- Jurisdictions establishing citizen oversight procedures or seeking to improve existing procedures need to consider four issues not discussed in detail elsewhere in this report:
  - Outreach.
  - Oversight structural considerations.
  - The openness of the procedures to the public.
  - The role of "politics."
- Effective outreach is essential to a successful oversight system; otherwise, allegations of misconduct will go unreported and the system will not be used.
- Despite its importance, most oversight bodies have lacked the resources to market their services effectively.
- Jurisdictions that engage in outreach:
  - Publish and distribute program brochures.
  - Place information about the system in the telephone book, police stations, and the mayor's office and on the Internet.
  - Promote coverage in the local press beyond attention to high-profile cases.
  - Give talks to neighborhood groups and other agencies.
  - Arrange for citizens to pick up complaint forms at multiple locations.
- Some citizens are reluctant to file complaints because they fear retaliation from the police.
- Jurisdictions need to address several organizational issues related to the structure of their oversight process, such as:
  - Developing the oversight system's legal basis (typically by municipal ordinance).
  - Determining which complaints will be investigated, reviewed, or audited.
  - Providing the system with subpoena power.
  - Minimizing delays in case processing.

> ## Key Points (continued)
>
> - Making oversight procedures public has potential benefits and drawbacks.
>
>   — Openness can increase the public's trust in the system but discourage citizens who want to remain anonymous from filing complaints.
>
>   — There are often legal barriers to opening citizen oversight to the public.
>
> - Most oversight bodies prepare annual reports for public dissemination. Some reports include:
>
>   — The nature and status of each policy recommendation the body made that year.
>
>   — Demographic information about complainants and subject officers.
>
>   — Cases in which the chief or sheriff disagreed with the board's findings.
>
> - "Politics" can seriously hamper the oversight system's effectiveness. Politics can involve:
>
>   — Conflict among local government officials.
>
>   — Volunteer board members with a pro-police or anti-police "agenda."

Jurisdictions setting up new citizen oversight procedures or improving an existing system need to consider four important issues that are not discussed in detail elsewhere in this report:

1. How to conduct effective outreach so that citizens know the oversight process is available to them.

2. How to structure the oversight process.

3. How public the system's procedures will be.

4. How "politics" can interfere with the system's effective operation.

Outreach

Citizen oversight bodies use a variety of methods to advertise their availability and services, but most have not done an effective job of publicizing themselves because:

- They lack the resources to market their services effectively.

- Local media tend to focus only on scandals related to police misconduct, not on mundane issues of how and why to file complaints.

- Police and sheriff's departments that take initial complaints may not make complainants aware of the citizen oversight option.

Nevertheless, outreach is important because, if citizens are not aware of the oversight body and its services, allegations of misconduct will go unreported. Reflecting this perception, in 1998 San Francisco's Office of Citizen Complaints (OCC) hired an additional staff person whose responsibilities specifically include community outreach, while Berkeley's Police Review Commission has established an outreach subcommittee.

To become widely known, oversight staff need to use multiple marketing approaches. As the following discussion and exhibit 5–1 suggest, many jurisdictions have implemented valuable outreach methods, but no jurisdiction has incorporated all of them.

Publicity materials

The Minneapolis Civilian Police Review Authority has published a brochure about its services in several languages. A map in San Jose's brochure identifies where the office and validated parking are located. San Francisco's citizen complaint form is formatted as a postage-paid, self-mailing letter (see appendix C). The Berkeley Police Review Commission (PRC) distributes

SOME CITIZENS FEAR OFFICERS MAY RETALIATE

Some citizens are afraid to report allegations of police misconduct because they fear officers will retaliate against them for complaining. One complainant reported, "I was concerned about retaliation—I felt if the officer could find out about this [complaint], I might want to rethink about whether to pursue the case. But the investigator said don't be concerned—he had had only one case of retaliation." Another complainant reported that he was waiting for someone in his car when the officer against whom he had filed his complaint rode by in a cruiser. After they made eye contact, the officer stopped and watched him. When the complainant's friend arrived, the officer drove off.

A complainant in one city expressed concern that, because the oversight office was located next to the police station, he was nervous that officers could see him enter and leave the building. As a result, oversight bodies try to locate their offices away from the police department. The Berkeley ordinance specifies that the board meetings "shall not be held in the building in which the Police Department is located."

However, oversight staff in most jurisdictions believe that actual retaliation is rare. Tucson's auditor has received only one complaint alleging retaliation, while San Francisco's Office of Citizen Complaints (OCC) received fewer than 10 complaints during the 3-year period between 1996 and 1999, none involving violence. OCC confirmed only one allegation.

Examination of data collected from citizen surveys and debriefings as part of the 1977 Police Services Study in Rochester, New York, St. Louis, Missouri, and Tampa-St. Petersburg, Florida, identified 455 individuals who felt they had a reason to complain about police conduct but took no action. A relatively small proportion of these citizens said they did not complain because they were afraid of the police (3.2 percent) or felt that filing a complaint would make matters worse (4.6 percent).

The most common reason for not complaining (42 percent) was the belief that filing would do no good.*

Oversight bodies can try to reduce retaliation—or the fear of retaliation—by telling complainants to report immediately any attempts at reprisal to the police department and the oversight board, where the allegation will receive prompt attention. San Francisco's OCC and Berkeley's Police Review Commission brochures inform potential complainants that retaliation is illegal. OCC staff also advise apprehensive would-be complainants to weigh whether they will be safer by complaining (and thus becoming known and identified) or by not complaining (and thus remaining vulnerable without any notice to those who could act to protect them). Police and sheriff's departments can reduce the chances of retaliation by developing and disseminating a clear policy prohibiting reprisals. A bulletin, the "Policy of the Police Commission of the San Francisco Police Department on OCC Cooperation," advises all members of the department of the following:

1. Attempts to threaten, intimidate, mislead, or harass potential or actual OCC complainants, witnesses, or staff members will be considered to be serious violations of General Order L–1 deserving of severe forms of discipline including, but not limited to, termination.

2. When the Chief of Police receives a sustained case involving a violation of General Order L–1, such case will be referred to the Police Commission for trial.

3. Members who are the subject of a complaint filed with the OCC shall not contact the complainant or witness regarding the issues of the complaint.

---
* Walker, Samuel, and Nanette Graham, "Citizen Complaints in Response to Police Misconduct: The Results of a Victimization Survey," *Police Quarterly* 1 (1) (1998): 65–89.

Exhibit 5–1. Oversight Outreach Methods

> Although most oversight bodies have had difficulty making the public aware of their existence and procedures, many have implemented parts of a comprehensive marketing strategy.
>
> Publicity materials:
> - Brochures (some in foreign languages).
> - Business cards.
>
> Postings:
> - Listings in the telephone directory.
> - Brochure and business card racks in the mayor's office.
> - An Internet site.
>
> Media:
> - Sending notices of hearings to the media.
> - Placing announcements in newspapers.
> - Televising hearings.
>
> Neighborhood groups and other agencies:
> - Mailing brochures and business cards.
> - Making presentations.
>
> Filing locations:
> - Providing filing forms at multiple locations.
> - Facilitating Internet filing.
>
> Referrals by police:
> - Posting signs in police stations.
> - Handing out oversight brochures and business cards.

a foldover business card describing PRC and the complaint filing process.

Postings

Most oversight bodies are listed in the telephone directory. However, because the agency's function may not be clear from its name, the public may not realize the agency is the place to contact to file a complaint against the police. Furthermore, even independent oversight bodies are sometimes listed in the phone book under "Police," which may discourage some citizens from filing complaints because they believe the organization is a part of the police department. San Francisco's Office of Citizen Complaints is listed in the business section of the phone book twice, once as Office of Citizen Complaints, San Francisco Police Department (boldface), with the police department address, and a second time with OCC's physical location.

The Minneapolis Civilian Police Review Authority's (CRA's) brochures are available, along with CRA business cards, in a wall display outside the mayor's office. The city's free events calendar lists the name and telephone number of the Civilian Police Review Authority under the "Police" heading.

Media

Tucson's auditor, as do other oversight directors, sends notices of each board agenda to newspapers, radio, and television. The Sunday newspaper lists the next council agenda. Some citizens learn of their jurisdiction's oversight body when the media cover a high-profile case of alleged police misconduct that involves the oversight system. A local cable station televises Portland's board meetings when appeals are heard.

Neighborhood groups and other agencies

Liana Perez, Tucson's auditor, sends pamphlets to community and neighborhood centers. She also alerts interested citizen groups to police issues, such as when she told the Southern Arizona People's Law Center that she was bringing up the issue of off-duty, uniformed officers working for merchants. Orange County Citizen Review Board members distribute their brochure when they give talks to civic groups.

Tucson's auditor uses the city's Citizen Neighborhood Services Department, a resource office for the city's 200 neighborhood associations, to send fliers to neighborhood associations offering to make presentations. OCC staff in San Francisco usually earn compensatory time to attend street fairs, community meetings, school assemblies, and other events while off duty to publicize the office. (See "San Francisco's Office of Citizen Complaints Monitors Selected Public Demonstrations.")

Filing locations

The more locations oversight bodies have in the community where citizens can pick up complaint forms, the easier it will be for individuals to file who may not have the time or assertiveness to travel to a central location. Citizens in San Francisco may file a complaint at any city agency, including the mayor's office and the sheriff's

SAN FRANCISCO'S OFFICE OF CITIZEN COMPLAINTS MONITORS SELECTED PUBLIC DEMONSTRATIONS

At the director's instruction, Office of Citizen Complaints (OCC) staff attend situations where there has been a public perception or actual history of police misconduct. Wearing OCC hats or jackets, they regularly monitor bicycle rallies, community fairs, and public demonstrations. A group sponsoring an Immigration Pride event wrote OCC asking staff to monitor the event because it claimed there had been problems with the police in the past; OCC agreed to go. No complaints resulted, and OCC staff did not observe any situations that required investigation of alleged misconduct.

OCC staff believe that the observations can serve to document, interpret, and evaluate the potential merits of OCC complaints, including establishing that there is no basis for a complaint.

The agency has developed a written policy for monitoring demonstrations (see appendix D) that specifies that "it is the policy of the OCC to monitor demonstrations when it is determined to be consistent with OCC's mission, and feasible and advisable to do so, in the joint determination of the Director and Chief Investigator."

At least one monitor and a supervisor observe each demonstration. Staff do not hand out intake forms and, except in emergencies, do not take complaints at the scene. If a civilian wishes to make a complaint, a monitor offers to provide the person with an OCC Incident Information Card (see exhibit 5–2) and, when appropriate, a business card. After gathering as much information as possible about the complaint, the monitor suggests the person contact OCC during business hours for followup.

---

department. Citizens in Omaha may pick up forms in any library branch. Melvin Sears, the Orange County Citizen Review Board administrative coordinator, set up a Web site citizens can access to get information about the board or to file a complaint online. The Web site on the city page set up by Liana Perez in Tucson also allows citizens to file complaints electronically.

Most oversight bodies lack the resources to set up satellite offices. According to Mary Dunlap, director of the Office of Citizen Complaints in San Francisco, "We should set up office hours in communities that are poor, young, and otherwise likely to underreport alleged police misconduct, like satellite mayor's offices, where we wouldn't take complaints but could explain the complaint process."

Referrals by the police

According to Mary Dunlap, "Every district station should also have a display at the window and a sign on the wall, along with brochures and complaint forms." A sign to the right of the Berkeley Police Department receptionist desk in fact says the following:

> If you have a complaint regarding a Berkeley Police Officer's conduct or need an explanation regarding a department practice, policy or procedure, you can either
>
> 1. Contact the Watch commander or Senior Officer in charge.
>
> 2. Contact the Internal Affairs Bureau of the Berkeley Police Department. The phone number . . . is 664–6653.
>
> 3. Contact the City of Berkeley Police Review Commission. The PRC is a civilian review board independent of the Berkeley Police Department. It is located at 2121 McKinley Avenue, phone 644–6716.

A brochure that Berkeley's internal affairs investigators give to complainants includes a four-paragraph description of PRC with its address and telephone number. The

The Web site on the city page set up by Liana Perez in Tucson also allows citizens to file complaints electronically.

Exhibit 5–2. OCC Incident Information Card

OCC INCIDENT INFORMATION CARD
THIS IS NOT AN OCC COMPLAINT FORM!

This card is to assist you in gathering the necessary information to file a complaint against a police officer. The OCC investigates complaints of misconduct of sworn members of the San Francisco Police Department. Complaints may involve allegations that a police officer has either acted improperly or has not properly performed a duty. You may file a complaint in person or by writing or calling the: Office of Citizen Complaints, 555-7th Street, San Francisco, CA 94103. Telephone: 415-553-1407.

Officer(s) Involved: (Badge No.'s, descriptions, car no.'s, etc.) _____

Time: _____ Date: _____
Location: _____
Witnesses: (Include phone no.'s if possible) _____

-NOTES-

BRING THIS INFORMATION WHEN YOU FILE YOUR COMPLAINT

Portland Police Bureau includes a notice about the review board with the letter it sends to complainants reporting their case findings (see exhibit 5–3).

Some Tucson beat officers hand out the independent police auditor's business cards. Officers who work the beat where the auditor is located periodically come in to ask for new supplies of cards, as do the secretaries at department substations.

Issues of Oversight Mechanics

Jurisdictions need to address—or reexamine—several organizational issues related to the structure of their oversight process.

> Tucson officers who work the beat where the auditor is located periodically come in to ask for new supplies of business cards, as do the secretaries at department substations.

Oversight's legal basis

Citizen review bodies have been established by municipal ordinance, State statute, voter referendum, mayoral executive order, police chief administrative order, and memorandum of understanding. The vast majority have been established by municipal ordinance.[1] Typically, the authorizing body or legislation grants the oversight body the power to adopt rules and regulations and develop procedures for its own activities and investigations. Examples of these rules include Berkeley's 16-page "Regulations for Handling Complaints Against Members of the Police Department" and Minneapolis' 28-page "Civilian Police Review Authority Administrative Rules." (See chapter 8, "Additional Sources of Help.")

Eligible complainants and cases

San Francisco accepts anonymous complaints but sustains them only with corroboration. Most jurisdictions permit aggrieved citizens and the parents of juveniles to file complaints. San Francisco also allows organizations to file complaints. The Berkeley Police Review Commission designated itself as the complainant in one case. Many oversight systems do not accept complaints by one officer against another officer.

Deciding what types of cases to review or investigate has important implications for staffing needs, system costs, and, above all, case processing delays (see "Minimizing Delays" on page 101). Mark Gissiner, past president of the International Association for Civilian Oversight of Law Enforcement, recommends, on the one hand, that systems not try to handle every type of complaint because the result can be a large backlog of cases whose resolution is delayed significantly, especially if hearings are held on each case. On the other hand, Gissiner says, oversight systems *should* investigate or review all cases involving use of firearms. Because most oversight systems have been established in response to an incident in which the police shot someone, the public expects an oversight body to review these cases. Indeed, the St. Paul and Orange County boards automatically review all cases in which an

Exhibit 5–3. Flier Included in Letter the Portland Police Bureau Sends to Complainants Notifying Them of Their Cases/Findings

---

NOTICE OF THE RIGHT TO APPEAL

If you are dissatisfied with these findings and conclusions, you may, not later than 30 days after the date of this letter, request review by the Police Internal Investigatons Auditing Committee (PIIAC). Failure to contact PIIAC within 30 days after the date of this letter will result in your loss of the right to request review.

You may reach PIIAC at:
Police Internal Investigations Auditing Committee
303 City Hall, 1220 SW Fifth Avenue
Portland OR 97204
(503) 823-4126

PIIAC accepts cases for review primarily to help the City refine police operations. PIIAC reviews whether the Bureau of Police conducted a thorough, timely and fair investigation of a complaint based on the evidence available to the police. The Police Internal Investigations Auditing Committee does not re-investigate complaints.

---

officer has discharged a firearm, even if there has been no complaint.

Jurisdictions must decide how long after the alleged misconduct occurred complainants may file a complaint. Berkeley's Police Review Commission requires complainants to file within 90 calendar days of the alleged misconduct, with another 90 days allowed if six board members vote that the complainant has demonstrated by clear and convincing evidence that his or her failure to file in time was the result of "inadvertence, mistake, surprise, or excusable neglect." Not knowing about PRC's existence or procedures does not fall into any of these categories. Furthermore, police testimony is not mandatory in cases that are accepted during the "late filing" period.

> The St. Paul and Orange County boards automatically review all cases in which an officer has discharged a firearm, even if there has been no complaint.

Subpoena power

As of 1995,[2] almost 40 percent of review bodies had subpoena power—the right to command an individual to appear to testify or produce documents—including oversight procedures in Berkeley, Flint, Orange County, Portland, and San Francisco. Legislation in Orange County provides for a fine of up to $500 and imprisonment for up to 60 days for officers who refuse to honor a board request to appear.

Most citizen oversight procedures that have subpoena power, including Berkeley's and San Francisco's, are prohibited from undertaking an investigation until any pending criminal charges against police officers have been adjudicated or unless they receive permission from the district attorney to proceed.

Many oversight advocates and directors believe that having subpoena power serves no useful purpose. If the oversight body already has authority under *Garrity* v. *New Jersey* (see "The Legality of Forcing Officers to Testify"), subpoena power adds nothing. If the oversight body lacks authority under *Garrity* to compel testimony, there is still little reason to seek subpoena power. Patricia Hughes, executive director of Minneapolis' Civilian Police Review Authority, and Daryl Lynn, CRA's chairperson, can remember only one case in which they could have benefited from having subpoena power. Indeed, the Orange County review board, Portland's city council acting as the Police Internal Investigations Auditing Committee, and the Flint ombudsman's office have never used their subpoena power. Furthermore, a 1992 city council report in Rochester suggested:

> The advantage of using the investigative authority of the IA lies in the fact that police officers are required to cooperate fully with the investigation since it falls within the employer-employee relationships. If the investigative authority were transferred to an outside agency, accused officers would be able to have recourse to their constitutional rights as citizens to avoid making any statement which might tend to incriminate them.

THE LEGALITY OF FORCING OFFICERS TO TESTIFY

Under *Garrity v. New Jersey*, 385 U.S. 493 (1967), whoever is the employer of a police officer, including not only the chief but, by extension, the city manager or mayor, can order the officer to answer questions "specifically, directly, and narrowly relating to performance of his [or her] official duties" as part of an internal, noncriminal investigation. Failure to answer questions related to the scope of their employment may form the basis for disciplining and dismissing officers. However, statements officers make under this requirement cannot be used against them in any subsequent criminal proceeding unless the officers are alleged to have committed perjury. Because officers can be terminated if they do not answer administrative questions, internal affairs and citizen oversight investigators typically read them their "*Garrity* rights"—a guarantee that the information sought will not be used against them in a criminal proceeding but that failure to respond to questioning could lead to disciplinary action.

In 1998 the Colorado Court of Appeals in *City and County of Denver and the Public Safety Review Commission* v. *Scott Blatnik and Jerome Powell* (97 CA 1662) held that law enforcement officers are entitled to assert their fifth amendment privilege before a citizen review board. The court ruled that the Denver Public Safety Review Commission could not compel subject officers to testify during an inquiry into allegations of improper use of force once the officers had invoked their privilege against self-incrimination. The city and commission had sued to compel them to testify after the officers had invoked the amendment. The court ruled that, because the commission was not the officers' employer, it could not compel them to testify.

The court of appeals distinguished this case from a 1997 Federal court case, *Pirrozzi* v. *New York* (950 F. Supp. 90 [S.D.N.Y. 1996], aff'd 117 F. 3d 7223 [2d Cir. 1997]), which compelled an officer to testify under threat of discharge. In *Pirrozzi*, the court found that officers can be compelled to testify as a condition of employment by employers *or those representing their employers*. Because New York City's review board is an integral part of the disciplinary process, and because departmental regulations require officers to give statements to the board under threat of termination, subject officers cannot invoke the protection of the fifth amendment.

The Berkeley city attorney issued a similar ruling in 1998. Because of the Colorado case, a police officer took the fifth amendment during an interview with the Police Review Commission investigator. As a result, Barbara Attard, the PRC officer, asked the city attorney for a legal opinion. The attorney ruled that, under a California statute similar to the Federal *Garrity* ruling, officers must testify because PRC acts pursuant to the authority of the city manager, who is the police department officers' employer.

Trying to secure subpoena power could involve oversight planners in lengthy court battles with officers' unions that they may not win. In addition, in the process of the litigation, planners may incur significant legal costs (see the second page of appendix E) and lasting poor relations with the police or sheriff's department. Although subpoena power could in some limited circumstances be useful for forcing citizens (e.g., complainants or, more likely, witnesses) to testify or provide documents, oversight staff are unlikely to want to exert such coercion.

Other structural issues

Jurisdictions that decide to have a citizen review board must settle other organizational and operational issues, some of which include the following:

- Should the entire board hear or review every case, as in Orange County and St. Paul, or should rotating groups of three or four members hear cases, as in Berkeley, Minneapolis, and Rochester?

- Should board members know what IA's findings are in advance of their own hearing? In Rochester, they do not.

- Legislation in Orange County and St. Paul permit boards to hire their own investigators if they are dissatisfied with internal affairs' investigations, but neither has ever done so. Although there are cost implications in hiring an investigator, the option may help motivate IA to do a better job with its own investigations.

- Where and when will hearings be held? As noted earlier, most jurisdictions try to house their oversight bodies some distance from the police station. (Because the St. Paul Police Department administers the oversight body, housing it in the public safety building is not an issue.) Most board hearings in Rochester take place during the day so the police department does not have to pay IA investigating sergeants overtime to attend evening meetings. Sometimes this creates a problem for employed board members who work a regular 9-to-5 day.

- Will the standard of evidence for sustaining a complaint be a preponderance of the evidence or the more stringent clear and convincing evidence? Some boards use one, some the other. Subject officers favor the more stringent standard, while complainants favor the more lenient standard.

- Finally, how can unacceptably long delays in reviewing and hearing cases be avoided? Delays are a problem for some oversight bodies.

Minimizing Delays

Many oversight bodies struggle to keep the review, hearing, or auditing process from taking months and even years to end. The annual report of the Tucson independent police auditor observes: "A concern that is frequently raised by complainants is the length of time taken to complete an investigation." Nearly two-thirds of complainants interviewed in a study of New York City's citizen oversight process reported the process took too long.[3] According to Jerry Sanders, former San Diego police chief, "Delays harmed the credibility of the review process here more than anything else." Sanders adds, "They also put officers under enormous stress" waiting for their cases to be decided.

> The Orange County review board, Portland's city council acting as the Police Internal Investigations Auditing Committee, and the Flint ombudsman's office have never used their subpoena power.

Delays were such a problem in San Francisco that the police commission directed the Office of Citizen Complaints to explain its backlog of cases. OCC's report, issued in February 1998, observed that most cases were completed within 1 year of receipt; when they were not, circumstances beyond OCC's control were often responsible, including the unavailability of participants or documents; delays requested by union representatives, criminal litigants, and attorneys; and staff attrition. The San Jose independent auditor's annual report includes a chart illustrating a sample of 10 cases and the number of days the complaint remained at different stages of the review process (see exhibit 5–4).

To reduce delays, many jurisdictions have established deadlines by which police departments and oversight bodies must complete their reviews.

CHAPTER 5: ADDRESSING IMPORTANT ISSUES IN CITIZEN OVERSIGHT

- The city council requires the Rochester Civilian Review Board to review cases within 2 weeks of IA's notification that its investigation is complete, but the board sometimes misses the deadline when a high-profile case takes precedence or it proves impossible to find three board members who can assemble within that 2-week period.

- The Minneapolis ordinance requires the Citizen Police Review Authority to complete a preliminary review within 30 days after a citizen has signed a complaint and complete an investigation within 120 days of the signing, with a 60-day extension allowed in rare circumstances. CRA has never missed a deadline because the police union then might argue to have the case dismissed.

- As of January 1, 1998, a new provision in the California Government Code requires the Office of Citizen Complaints to conclude investigations of complaints and make findings within a year of filing, absent exceptional circumstances.

Some oversight bodies establish internal rules for completing cases. However, according to one activist, "Establishing hard deadlines without adequate money for staff is a setup for failure."

> *Patricia Hughes, executive director of Minneapolis' Citizen Police Review Authority, came up with the idea of plea bargaining cases through a stipulation procedure that decreased the need for hearings dramatically, thereby reducing delays for other cases.*

Berkeley helped reduce its backlog of cases by amending its regulations so that it would not be required to hold a hearing on every filed complaint, instead allowing the director to recommend that the Police Review Commission summarily dismiss cases without merit. Patricia Hughes, executive director of Minneapolis' Citizen Police Review Authority, came up with the idea of plea bargaining cases through a stipulation procedure that decreased the need for hearings dramatically, thereby reducing delays for other cases. (See the Minneapolis case study in chapter 2.) A citizen advisers' monitoring report to the Portland City Council highlighted delays in processing complaints at the police bureau's internal affairs department and provided four strategies for reducing the delay, including improved recruitment and staffing and the establishment of timeliness goals for each stage in the review process.

When a delay is inevitable, Felicia Davis, administrator of the Syracuse, New York, Citizen Review Board, sends the complainant a letter explaining where the case is in the complaint process and the reasons for the delay. Davis says, "In effect, I tell them, 'We haven't forgotten you.' This helps keep them interested in and willing to pursue the case." Davis sends the letter after 60 days when she knows the case will take more than 90 days to be decided.

EXHIBIT 5–4. NUMBER OF DAYS EACH OF 10 COMPLAINTS REMAINED AT 3 SAN JOSE POLICE DEPARTMENT OFFICES

Case	IA	Other Bureau	Chief	Total Length of Investigation and Administrative Review (days)
1	582	259	37	878
2	591	57	68	716
3	310	118	105	533
4	230	154	64	448
5	56	41	342	439
6	94	177	163	434
7	176	259	173	608
8	228	136	181	545
9	43	301	139	483
10	125	74	152	351

Openness of Oversight Proceedings

Most citizen oversight advocates feel strongly that oversight proceedings benefit from openness.

> Many experts believe that one of the most important functions of citizen oversight is to provide information to the public about the police department and the complaint process. By itself, this information serves as a form of oversight and accountability, providing voters, elected officials, and the news media with relevant information about police activities. Information serves to "open" police departments to the public.[4]

However, oversight procedures have to be sensitive to the legal and ethical privacy rights of complainants and police officers. For example, State public records, statutes, and labor-management agreements may limit the information citizen oversight bodies can disseminate to complainants and the public.

- A California statute provides that "Peace officer personnel records and records maintained by any state or local agency, . . . or information obtained from those records, are confidential and shall not be disclosed in any criminal or civil proceeding except by discovery." Disclosure, with narrow exceptions, is a criminal offense.

- The corporation counsel to the city of Rochester advised that "the public airing by the Police Advisory Board [sic] of its agreement or disagreement with the findings of the Chief of Police would not appear to be fully in keeping with the intent of the New York State's Civil Rights Law, which makes all personnel records used to evaluate performance toward continued employment or promotion of a police officer confidential."

Even when not prohibited, openness and access may discourage citizens who want complaints kept secret from coming forward, and it may inhibit officers from reporting misconduct by other police personnel.

When legal, activities that oversight procedures can consider making public include:

- Hearings.
- Findings.
- Policy recommendations.
- Internal quality control findings.

Hearings in Berkeley and Orange County are open to the public. The Berkeley ordinance requires that all commission meetings and agendas be publicized at least 3 days in advance by written notice to newspapers, radio, and television stations serving the city. The Orange County board invites 57 media outlets to board meetings. Flint's ombudsman's office faxes its findings to the local newspaper, two radio stations, and three television stations.

The Orange County board invites 57 media outlets to board meetings.

Many oversight bodies write complainants about the outcomes of their cases but, because of legal limitations or the chief's decision, rarely report what discipline was imposed—and sometimes whether discipline was imposed.

Finally, all oversight bodies prepare annual reports (and sometimes monthly or quarterly reports). (See "The San Jose Office of the Independent Police Auditor Annual Report Is Particularly Informative.") At a minimum, these reports should include:

- The disposition of complaints.
- Patterns of complaints, such as:
  — Type.
  — Geographic area.
  — Race, ethnicity, and gender of complainants.
  — Characteristics of the officers (e.g., race, gender, assignment seniority).
- Any policy recommendations.

THE SAN JOSE OFFICE OF THE INDEPENDENT POLICE AUDITOR ANNUAL REPORT IS PARTICULARLY INFORMATIVE

The *1997 Year End Report of the San Jose Office of the Independent Police Auditor* is a particularly comprehensive and well-presented document. The 58-page, spiral-bound report, with a glossy burgundy cover, includes:

- Biographical sketches of office staff.

- An 11-page executive summary printed on burgundy-colored pages.

- A flowchart illustrating the complaint process.

- A discussion of complaint timeliness that includes a chart illustrating a sample of 10 cases and the number of days a complaint remained at different stages of the review process (see exhibit 5–4).

- The types of complaints and sustained cases by city council district for the previous 3 years.

- A chart showing the type of alleged unnecessary force used by body area affected and degree of injury.

- Demographic information about complainants, including gender, ethnicity, age, educational level, and occupation.

- Statistical information about subject officers, including bureau, gender, years of experience, type of allegation by years of experience, and police unit in which they work(ed).

- A chart showing discipline imposed.

- A discussion of the criteria for evaluating internal affairs investigations and the auditor's findings related to each criterion.

- Summaries of seven selected audited cases.

- A chart showing the status of every policy recommendation the auditor has presented and its disposition since the office was established in 1993.

Politics

"Politics" may interfere in two respects with the effective operation of citizen oversight.

Conflict among local government officials

Conflict among elected and appointed officials in a jurisdiction over the operation of citizen oversight can disrupt the review process. In Flint, the mayor appoints the police chief, but the city council appoints the ombudsman. If the mayor and council do not see eye to eye, there is the potential for conflict with the ombudsman becoming the chief's adversary. Even when there is a good working relationship among the involved officials in a jurisdiction, turnover through new elections or appointments can result in new personnel who wish to have things done "their way."

Sometimes, ambiguity in the lines of authority and communication creates the potential for controversy:

- The Portland mayor appoints and hires the auditor, but the auditor feels she is legally responsible to the board—which is the city council sitting en bloc— because she acts as the board's executive director. Who her true legal supervisor is has never been tested. The auditor is also in an awkward position whenever her reports are critical of the city because what she says could make the city liable for damages. As a result, there were times when the mayor expressed concern about the auditor's statements.

- The Berkeley Police Review Commission officer is staff to the city manager, who appoints her, but the perception among the public and police department is that she is staff to PRC. The city manager in effect delegates his oversight role to the PRC officer. As a result, the officer needs to maintain good relations with both PRC and the city manager.

In one jurisdiction, board members do not consult with the council members who appointed them when reviewing specific cases. However, because some council members tend to side with the police, and others are hostile to the department, they appoint board members sympathetic to their respective positions. As a result, the board is split between pro- and anti-police factions.

Politics also can work for the good. In some jurisdictions, the police chief cooperates with the oversight process at least in part because the mayor demands support for it. Because the Minneapolis city council president led the effort to revamp the city's oversight system before she became president, all the involved parties understand that she expects them to cooperate.

Agendas on the part of volunteers

When they are appointed by the mayor or council members, volunteer board members may feel, as Lt. Robert Skomra, former head of IA in Minneapolis, pointed out, that "they represent a special interest and see themselves as champions for that group." As one city council member said, "Some board members play to the tune of the city council member who appoints them." In one jurisdiction, some board members have supported the electoral campaigns of the council members who appointed them; these volunteers may feel especially obligated to reflect "their" council member's political views.

> *Of course, not all disagreement among elected and appointed officials and not all bias among oversight volunteers is politically motivated in the sense of serving narrow self-interests. Many clashes over citizen oversight are the result of genuine differences of opinion on what is, after all, a controversial topic.*

A partial solution to this problem of volunteer bias may be mediation training. Volunteer board members in Rochester all must become certified mediators, which may increase their ability to provide impartial reviews (see chapter 4, "Staffing"). Another approach to ensuring board members' objectivity is to avoid having government officials select them. Each of seven neighborhood coalitions recommends one individual adviser to serve as a citizen adviser in Portland. Board members in Rochester are selected by the review board's screening committee, consisting of board chairpersons and staff. As a result, according to Anne Pokras, former director of special projects for the board's parent agency, "Panelists have a heightened awareness that they represent no one—that is, no politician—but everyone—that is, the community."

Some Portland residents have called for the election of board members. However, according to a local activist, this might result in the police union's providing more campaign funds to candidates sympathetic to their positions than other candidates could raise (see "Working With the Union" in chapter 6).

Of course, not all disagreement among elected and appointed officials and not all bias among oversight volunteers is politically motivated in the sense of serving narrow self-interests. Many clashes over citizen oversight are the result of genuine differences of opinion on what is, after all, a controversial topic. Chapter 6 identifies some of these conflicts as they relate to oversight bodies and police and sheriff's departments.

Notes

1. Walker, Samuel, *Citizen Review Resource Manual,* Washington, D.C.: Police Executive Research Forum, 1995: 6–7.

2. Ibid., 13.

3. Sviridoff, Michele, and Jerome E. McElroy, *Processing Complaints Against Police in New York City: The Complainant's Perspective,* Washington, DC: Vera Institute of Justice, 1989: 109.

4. Luna, Eileen, and Samuel Walker, *A Report on the Oversight Mechanisms of the Albuquerque Police Department,* prepared for the Albuquerque City Council, 1997: 131.

Chapter 6: Resolving Potential Conflicts Between Oversight Bodies and Police

KEY POINTS

- Three preliminary steps can help significantly to reduce conflict among all the parties involved in citizen oversight:

  — Either initiate the oversight system without the impetus of a controversial police shooting or avoid consideration of the incident in the planning process.

  — Involve representatives of all concerned parties in the planning of the oversight procedure.

  — Establish clear, measurable objectives for the oversight system.

- Many police administrators and officers have criticisms of local oversight bodies, most of which fall into three categories:

  — Citizens should not interfere with police work.

  — Citizens do not understand police work.

  — The process is unfair.

- Several considerations and actions can help address police concerns about the oversight process, including:

  — Recognizing the typically advisory role oversight bodies play but also documenting the judicious role most oversight systems have adopted.

  — Training board members thoroughly and publicizing how carefully they have been prepared.

  — Accepting that the mission of oversight *is* to provide for citizen, not professional, review.

  — Highlighting that oversight bodies agree with the police or sheriff's department's findings in the vast majority of cases.

  — Publicizing particularly high-profile cases in which the oversight body has sided with the subject officer(s).

  — Working to reduce delays in holding hearings and reviews.

  — Explaining how oversight findings can benefit officers.

  — Sitting down and resolving misconceptions and conflicts face to face.

> **KEY POINTS (CONTINUED)**
>
> - Oversight staff often have criticisms of the police. Their most common concerns are:
>
>   — Officers may refuse to answer questions, and departments may refuse to share records.
>
>   — Officers do not understand the oversight body's mission and legitimacy.
>
>   — Departments ignore the oversight body's findings or policy and procedure recommendations.
>
> - Some police departments have attempted to work constructively with their local oversight bodies, including disciplining officers who fail to appear for questioning and arranging for oversight staff to explain their procedures to officers at the academy or at roll call.
>
> - Oversight planners and review bodies need to take the initiative to involve union leaders in their activities. Some unions no longer oppose citizen oversight as strongly as in the past. Oversight planners successfully have:
>
>   — Involved union leaders in designing and setting up the review procedure.
>
>   — Accommodated some union concerns.
>
>   — Addressed union concerns about biased review procedures by ensuring the review process is scrupulously fair.
>
>   — Highlighted shared objectives, such as a joint interest in fair treatment of officers by internal affairs.

There are conflicts in many jurisdictions between oversight bodies and police agencies. To be sure, if there is *no* tension between them, the oversight body may not be acting assertively to maintain or improve police accountability. However, excessive conflict will destroy any oversight system.

Preliminary Steps for Minimizing Conflict

There are three preliminary steps jurisdictions can take that can substantially reduce the potential for future conflict not only between the oversight body and the police but among all parties involved in, or at least concerned about, citizen oversight—local public and elected officials, union leaders, and community activists.

Do not wait for a serious incident, typically a police shooting that creates a public uproar, before setting up an oversight system.

Involve representatives from all concerned parties as colleagues in the planning process.

1. Do not wait for a serious incident, typically a police shooting that creates a public uproar, before setting up an oversight system. Because of the tensions such an incident creates, it is difficult for the parties involved to approach the planning task in a rational manner. As a result, the planning process may perpetuate, rather than defuse, existing tensions. If the planning process has begun only after a conflict, avoid discussion of the incident in the planning process.

2. Involve representatives from all concerned parties as colleagues in the planning process. Although it may require months to iron out differences, even if they are not resolved to everyone's satisfaction, the implementation and operation of the oversight procedure is likely to proceed more smoothly if all the parties have participated in its planning. (See "Working With Activists.")

WORKING WITH ACTIVISTS

Local activists have often been as critical of oversight systems as have police departments and unions. Sometimes they criticize the system's lack of power; other times, they report that oversight staff are not using the authority they have to pursue cases of alleged police misconduct. One activist observed, "San Francisco's citizen oversight organization has the most money and best structure in the Nation, yet it sustains only 10 percent of cases."

A newsletter published by Dan Handelman, a member of Portland Copwatch, an organization that tracks alleged police misconduct, objects to the fact that the Portland police chief can ignore, and in the newsletter's opinion has ignored, reversals of IA findings by the city council acting in its capacity as the city's oversight board. However, in one of the two recent examples when the chief did this, the council's vote to reverse was 3 to 2, suggesting that there was room for honest disagreement. As a result, the chief's decision in this case, although it rejected the council's decision, was not necessarily arbitrary. However, the council's vote in the second matter was 4 to 1. In any case, Handelman's larger concern is that the city council—that is, elected citizens—not the chief, should have the final say in determining whether officers engaged in misconduct.

According to a board member in another city, "Some groups are very vocal and bring police problems to the media and raise holy hell. But if they didn't, we would not have achieved this level of oversight. So they play a beneficial role, but they can make life painful because they say some ridiculous things."

3. Specify precisely the review system's objectives. Without specific objectives such as the ones listed in chapter 7, "Monitoring, Evaluation, and Funding," the involved parties may lock horns because they have different expectations of what the system should be doing and accomplishing. Even if all the involved parties do not agree on what the oversight system should be trying to accomplish, at least they will have the same understanding of its goals.

That said, it remains true that the most severe antagonism surrounding citizen oversight is usually between the oversight body and the police or sheriff's department and between the review process and union leaders. This chapter reviews some of the principal sources of conflict between oversight bodies and law enforcement agencies—including police unions—and suggests possible solutions.

Police Criticisms of Oversight Procedures

Exhibit 6–1 summarizes the concerns many police and sheriff's departments express about citizen oversight along with possible responses to their concerns. As shown in the exhibit and discussed below, these concerns generally fall into three categories:

1. Oversight procedures represent outside interference.

2. Oversight staff lack experience with and understanding of police work.

3. The oversight process is unfair.

Citizens should not interfere with police work

Most police administrators believe their agencies should have the final—and often only—say in matters of discipline, policies and procedures, and training. Police administrators feel they have to be held accountable for their officers' behavior because they are in charge. Without final say over discipline, policy, and training, their accountability is undermined. (See "Should Citizens Control the Discipline Process?")

Police executives' objections to citizen oversight sometimes reflect their belief that they already do a good job responding to citizen complaints. As a result, when a finding from an oversight body disagrees with the department's internal finding, some chiefs and sheriffs

Exhibit 6–1. Concerns Many Police and Sheriff's Departments—and Union Leaders—Express About Citizen Oversight—and Possible Responses

Assertion: Citizens Should Not Interfere in Police Work

Concerns	Responses
• The chief must be held accountable for discipline to prevent misconduct.	• Most oversight bodies are only advisory.
• Internal affairs already does a good job.	• Even when the department already imposes appropriate discipline without citizen review, an oversight procedure can reassure skeptical citizens that the agency is doing its job in this respect. • The next chief or sheriff may not be as conscientious about ensuring that the department investigates complaints fairly and thoroughly.

Assertion: Citizens Do Not Understand Police Work

Concerns	Responses
• Oversight staff lack experience in police work.	• Board members typically have pertinent materials available for review, and ranking officers are usually present during hearings to explain department procedures. • Oversight administrators need to describe the often extensive training they and their staff receive. • Citizen review is just that—*citizens* reviewing police behavior as private citizens.
• Only physicians review doctors, and only attorneys review lawyers.	• Doctors and lawyers have been criticized for doing a poor job of monitoring *their* colleagues' behavior.

Assertion: The Process Is Unfair

Concerns	Responses
• Oversight staff may have an "agenda"—they are biased against the police.	• Oversight staff need to inform the department when they decide in officers' favor. • Oversight staff and police need to meet to iron out misconceptions and conflict.
• Not sustained findings remain in officers' files.	• Indecisive findings are unfair to both parties and should therefore be reduced in favor of unfounded, exonerated, or sustained findings.
• Adding allegations unrelated to the citizen's complaint is unfair.	• Internal affairs units themselves add allegations in some departments.
• Some citizens use the system to prepare for civil suits.	• Board findings can sometimes help officers and departments defend against civil suits.

SHOULD CITIZENS CONTROL THE DISCIPLINE PROCESS?

Most oversight directors and researchers agree that citizens should not have power to discipline officers. They believe that giving citizen oversight systems that authority would be illegal or unwise because:

- It would violate State law, city charter, or collective bargaining agreements with police unions.

- It would detract from holding the chief or sheriff accountable for ensuring proper standards of professional conduct, making it possible for the top law enforcement executive to argue, "Yes, my department has a problem with police misconduct, but I can't do anything about it."*

---

* Luna, Eileen, and Samuel Walker, *A Report on the Oversight Mechanisms of the Albuquerque Police Department*, prepared for the Albuquerque City Council, 1997: 148.

give little or no weight to the oversight body's finding in determining discipline. In some jurisdictions, chiefs and sheriffs accept internal affairs findings and decide on discipline long before they even receive the oversight body's findings. Some have never changed an IA finding as a result of an oversight finding that was different.

Jurisdictions have used a variety of strategies to address concerns about outside involvement in police matters. In most jurisdictions, local government has established oversight bodies that are only advisory; their recommendations are nonbinding on departments. Some review bodies can appeal the chief's or sheriff's rejection of their recommendations to elected or appointed officials who can require the department to act. However, because these officials have this authority regardless of whether there is an oversight body, the oversight procedure itself does not further diminish the police or sheriff's department's authority. Even when citizen oversight bodies do have some authority over the police, they have generally exercised it cautiously. For example, Flint and St. Paul have never used their subpoena power to compel officers to testify; Orange County and St. Paul have never exercised their right to hire an independent investigator to second-guess an IA investigation.

Many internal affairs and other procedures for investigating citizen complaints are already rigorous. However, the process typically is only as effective as the current chief or sheriff requires it to be. Advocates of citizen oversight believe it is important to have an independent review mechanism in place that can help IA maintain its standards in case the next chief or sheriff fails to demand—and ensure—fairness and thoroughness in the internal complaint investigations process.

> *Advocates of citizen oversight believe it is important to have an independent review mechanism in place that can help IA maintain its standards in case the next chief or sheriff fails to demand—and ensure—fairness and thoroughness in the internal complaint investigations process.*

In addition, even with a conscientious chief or sheriff, because of turnover there are usually some IA investigators who are not yet fully skilled in their jobs—and who may switch assignments after they have become fully qualified. Finally, even when a police or sheriff's department is being conscientious in imposing appropriate discipline without citizen review, an oversight procedure can reassure skeptical citizens that the agency is indeed following through responsibly on citizen complaints.

Citizens do not understand police work

Some police oppose citizen oversight procedures because they believe that oversight staff, lacking experience as police officers or sheriff's deputies, cannot determine fairly whether officers have engaged in misconduct. Officers frequently observe that State medical boards composed only of physicians investigate doctors for malpractice, and only attorneys investigate lawyers for misconduct. Similarly, some police argue, only law enforcement officers have the knowledge to investigate and judge other sworn personnel.

There are at least four specific areas in which police officers feel citizens do not understand police work:

1. Case law (and the agency's own rules) governing police behavior, such as when officers may conduct searches (and what types of searches) and when they may use deadly force against a fleeing felon.

2. The nature of police discretion, both in terms of what officers have the flexibility to do and what they may not do.

3. How officers are trained, in light of the fact that, in the absence of a policy or procedure, training *is* policy.

4. The manner in which the totality of the circumstances influences an officer's behavior—for example, when courtesy is not always a viable option.[1]

Several considerations may reduce these concerns:

- By reviewing materials they receive before or at each hearing, which typically include the work product of the IA investigation as well as relevant department policies and procedures, board members report they generally can determine whether officers engaged in misconduct. In addition, a department supervisor attends hearings in many jurisdictions (e.g., Berkeley, Orange County, and Tucson) or is available on call (e.g., Rochester) to answer questions about department operations.

- Regarding board members' lack of expertise, Jackie DeBose, a 10-year member of Berkeley's Police Review Commission, observes, "This is a *citizens'* review, not a court of law, so they [board members] *should* look at the problem as private citizens." In addition, Sgt. George Cardenas, the only sworn member of Omaha's Citizens Complaint Review Board (CCRB), notes, "In looking at whether officers violated a policy or procedure, board members are pretty good at determining the answer. But most cases [in his jurisdiction] are of the 'he said/she said' variety, so they don't need special expertise."

- Although selecting only individuals with police experience for board membership would negate the purpose of citizen oversight, citizen review systems that investigate allegations of misconduct can hire investigators with pertinent law enforcement expertise. Most of the Minneapolis Civilian Police Review Authority's investigators are former police officers who worked in other jurisdictions. Of course, these investigators need to be screened, trained, and supervised closely to make sure they do not show bias in favor of subject officers during their investigations.

- Although it is generally true that only physicians and attorneys investigate their respective colleagues for misconduct, many organizations and individuals have in fact criticized these licensing boards for ineffectively monitoring and disciplining members of *their* professions.[2] It is also notoriously difficult to find physicians who will testify in court against other physicians and lawyers who will testify against other lawyers. As a result, the analogy with only police overseeing other police is a poor one. If anything, the analogy reinforces the case for *not* leaving oversight to members of the profession being monitored.[3] Furthermore, as Mary Dunlap, director of San Francisco's Office of Citizen Complaints, points out, "Lay jurors already factually resolve allegations of police misconduct [and physician and lawyer malpractice] in civil and criminal justice trials that are a key element of the American justice system."

- Oversight directors need to educate officers about what their staffs do as well as to describe their backgrounds and training. According to Todd Samolis, coordinator of the Rochester Civilian Review Board, "At a panel for middle school students that I ran jointly with 10 officers, the officers were stunned to learn about the extensiveness of the mediation and academy training board members receive, especially impartiality training." Oversight staff in Rochester and Minneapolis attend roll calls so they can describe their operations and training to officers.

The process is unfair

Although many officers believe that the oversight process is unfair because, as previously discussed, citizen reviewers are unfamiliar with police work, officers also find the process unfair for other reasons. Many officers feel oversight staff have "an agenda"—that is, the staff believe it is their personal mission or assignment from the elected officials who appointed them to reduce police officers' power. As one chief said, "The problem is not the concept [of citizen oversight] but using biased staff. . . . [For example,] staff ask complainants exculpatory questions, but they don't ask exculpatory questions of officers that would justify *their* behavior." According to a report evaluating Tucson's oversight system, "The appointment of [board] members with strong political agendas can result in their use of the review body as a tool for promoting those causes."[4]

When they are biased, investigators and board members can sometimes come across as hostile. An officer who was a witness at a board meeting felt the process was so demeaning and insulting, she wrote a memo to the chief saying she would walk out of any future hearings: "The board chairperson was argumentative and condescending, and he in effect accused me of lying." Bias, some officers feel, also leads staff to tolerate complainants who are only out to get the police, such as drug dealers who regularly file complaints only to build a specious defense on grounds of harassment and to slow down assertive officers by getting complaints into their personnel folders.

Officers in several jurisdictions reported that some complainants take advantage of the complaint process to benefit a planned or ongoing civil suit against the city or officer. Because everything said at hearings in some jurisdictions is discoverable, plaintiffs in civil cases in effect get free "prediscovery." Daryl Lynn, chairperson of the Minneapolis Civilian Police Review Authority, reports that "some [citizens] do file complaints just to get discovery of the police department's case to use in a civil suit." After Mark Gissiner, a Cincinnati human resources analyst, conducted an investigation and prepared a report that concluded that an officer had used excessive force, a lawyer used the report to sue the city for damages. The city settled for $300,000. A plaintiff's attorney in another jurisdiction described how he makes use of the oversight process (see "An Attorney Uses Citizen Oversight as a Screening Tool for Civil Suits").

Officers in several jurisdictions felt that the practice of "added allegations," also called "collateral misconduct," was particularly unfair. Some oversight procedures, including those in Berkeley, Flint, Rochester, and San Francisco, investigate and sustain allegations that are not part of the citizens' original complaint but that oversight staff discover in the course of their investigation or review. For example, the complaint may be about the alleged use of foul language, but the oversight body learns the officer was not wearing a badge at the time of the incident or failed to file a report. One oversight body sustained an added allegation against an officer for not "timing out"—recording the end mileage on a wagon transport. These added allegations are not "citizen" complaints but accusations of misconduct by the review body.

Officers complain that they are held accountable for tiny rule violations, such as placing the wrong offense code on a citation. One chief said, "It gets tiresome to get nitpicked." Officers also object to having "not sustained" findings included in their file. Finally, many officers complain about the long delays the complaint process

> *Officers in several jurisdictions felt that the practice of "added allegations" was particularly unfair.*

AN ATTORNEY USES CITIZEN OVERSIGHT AS A SCREENING TOOL FOR CIVIL SUITS

A private attorney reported that he has used sustained citizen oversight cases as evidence in civil suits, and he may let the oversight agency do its investigation first so he can benefit from its work product when only a small settlement will be involved. When the board does not sustain a case, he reexamines the strength of his case. As a result, he uses the oversight system as a screening device. However, he has still won cases the oversight body has rejected, but he also lost a case the board sustained—even though there was a lower burden of proof in the civil case (preponderance of evidence) than in the board hearing (clear and convincing).

sometimes entails. According to one, "Delays make it impossible for officers to collect witnesses or even remember what happened."

Several actions and observations may help to temper police concerns that oversight procedures are unfair.

- Every citizen oversight process should protect officers from petty and vengeful complaints. "An important component of the intake process surrounds development of procedures (and related training) to assess and dismiss complaints that are unfounded . . . to permit frivolous complaints to be set aside and not utilize [oversight] resources unnecessarily" or inappropriately implicate and take up the time of officers.[5]

- Oversight boards need to let officers know when they decide in favor of officers in specific cases so police do not develop or maintain the misperception that the program is biased. (See "The Types of Board Findings Oversight Staff Should Publicize to Officers.") Boards can also make clear that they sustain citizen complaints at low rates that are not significantly higher—and sometimes lower—than those of internal affairs units.[6] Lisa Botsko, Portland's first auditor, went to training sessions for IA investigators to tell them how well they were doing.

- Investigators and board members should ask neutral questions in a nonaccusatory manner. One accused officer was pleased to report that the Tucson auditor's questions during an IA interrogation of his alleged misconduct were not judgmental but rather were designed "to get a clear picture of what happened." He felt her questions "all made sense."

- Mary Dunlap, director of San Francisco's Office of Citizen Complaints, observes that indecisive findings can be unsatisfying and unfair to *both* parties. As a result, she wants to reduce the number of undecided cases—that is, have more determinations that the case is unfounded, exonerated, or sustained. In any case, it may be the police or sheriff's department decision, not the oversight body's determination, to record unfounded cases in an officer's files.

- Reasonable doubt in a criminal case results in a "not guilty," not an "innocent," finding. It is perhaps not unreasonable, therefore, that in oversight cases, when a preponderance of the evidence is lacking that the officer did not engage in misconduct, an unsustained—not an unfounded or exonerated—finding is made. Similarly, just as a criminal defendant found not guilty still has an arrest record, it may not be unreasonable to include a record of unsustained cases in an officer's personnel

THE TYPES OF BOARD FINDINGS OVERSIGHT STAFF SHOULD PUBLICIZE TO OFFICERS

A mayor pointed out, "Even though review boards side with subject officers in the overwhelming number of cases, many officers still believe the board is out to get them." As a result, it becomes especially important to let officers know how specific reviews have favored them.

An Orange County sheriff's deputy wiped pepper spray on his hand and then wiped it on an unconscious suspect to wake him up. The IA unit sustained the violation of the department's pepper spray policy—excessive use of force—but the board said he was a new deputy from another department where deputies had carried ammonia capsules, and he was only using the spray as a substitute. The board said the sheriff's department pepper spray policy needed revision because it required automatic termination for misuse regardless of mitigating circumstances. As a result, the department rewrote its policy so that misuse would not require automatic termination. The deputy was suspended but not terminated.

When three Office of Citizen Complaints (OCC) staff monitored New Year's evening partying, they observed that San Francisco police officers used the utmost restraint in preventing a riot by drunken revelers. OCC issued an oral report to the police commission describing the officers' exemplary behavior.

file. A reasonable compromise may be for the oversight body not to report unsustained findings to the police or sheriff's department or for the department not to include them in the officer's file until the number of unsustained findings has reached an agreed-upon minimum number during a specified period of time (e.g., three findings during a 2-year period).

- Police IA units themselves add allegations to citizens'—or their own—complaints against officers. The Rochester Police Department's IA unit added what it calls 12 "satellite issues" as a result of investigations initiated for other reasons in 1997. A letter the Tucson Police Department sent to one complainant noted, "This complaint has been closed as OTHER, meaning that the Officer committed a violation of TPD [Tucson Police Department] Rules and Procedures other than the alleged violation." In other words, in adding allegations, citizen oversight bodies are merely following in police footsteps. In some jurisdictions, such as San Francisco, oversight bodies are required to investigate any added allegations they discover because their charter mandates that they investigate any wrongdoing they uncover. Nevertheless, San Francisco's internal affairs unit is trying to work with the Office of Citizen Complaints on not generating added allegations that do not represent deliberate misconduct—for example, claiming an officer recorded the wrong code on a report. Finally, on the one hand, added allegations are not citizen complaints and therefore, absent a legal mandate to investigate them, might be considered beyond the oversight body's purview. On the other hand, few citizens are familiar with their police or sheriff's department general orders and therefore are unaware when an officer's behavior violates these orders. A possible reconciliation of this dilemma may be for oversight bodies to ask the complainant if *he* or *she* would like to add any instances of misconduct discovered by the oversight body to the original complaint. If the complainant does not wish to add them, the oversight body could inform the department of the additional violations without including them as part of the complaint.

- Board decisions can benefit, not just harm, officers who are sued civilly. In one case, a citizen whose complaint a board did not sustain filed a civil suit, and the city attorney had the oversight investigator testify. The investigator's testimony helped the officer have the suit dismissed. (See chapter 1, "Introduction," for further evidence that citizen oversight can prevent or reduce award amounts in civil suits.) Whenever the department is sued, the Portland Police Bureau internal affairs unit shares with loss-control personnel reports that the Portland auditor routinely sends it in case the information can help the department's case.

- Delay is a product of many people. Also, slowness is often an unavoidable feature of many oversight bodies because they are overworked and understaffed, just as it can take years for Federal and State equal opportunity employment commissions to hear cases. Providing citizen oversight bodies with adequate personnel and funding often can reduce delays dramatically, as happened when San Francisco funded additional investigators for the Office of Citizen Complaints. Oversight bodies, such as Berkeley's, also can develop a process for summarily dismissing inappropriate complaints to reduce delays involved in scheduling hearings on legitimate complaints. Orange County allows board members to place noncontroversial complaints on a "consent agenda," bypassing the need for further discussion.

> *Oversight staff and police need to meet and talk, whether to iron out specific misconceptions or conflicts or to share information about what they are doing.*

- Oversight staff and police need to meet and talk, whether to iron out specific misconceptions or conflicts or to share information about what they are doing. According to Prentice Sanders, assistant chief of the San Francisco Police Department, in 1997 the police commission chairperson invited OCC and the police department as well as interested citizen groups to four roundtable discussions run by two independent professional facilitators to address concerns about the oversight process. Everyone looked for common ground. All agreed there was a problem with cases taking too long that needed to be addressed. They also agreed to talk to each other in the future if a problem arose before going to the press.

Oversight Criticisms of the Police

Oversight staff raise several concerns about their relationships with the police and sheriff's departments whose investigations they review. Their most common concerns are that in some cases:

- Officers refuse to answer questions.

- Departments resist providing needed records—or fail to provide them in a timely manner.

- Officers do not understand the oversight system's mission and legitimacy.

- Departments ignore the oversight body's findings or policy and procedure and training recommendations.

For example, the Berkeley Police Review Commission was trying to have the police department provide it with the relevant police report and computer-aided dispatch printout regarding radio and telephone communications before, not after, the commission gives a copy of the citizen's complaint to the police department. The Omaha Police Department was providing the Citizens Complaint Review Board with the results of investigations 59 days after receipt of complaints, leaving the board 1 day in which to review the findings according to the ordinance. The mayor issued an executive order requiring submission within 30 days or granting an automatic extension for the board's review. The minutes of the January 1998 Police Internal Investigations Auditing Committee (PIIAC) meeting in Portland record a citizen adviser as reporting:

> An ad hoc subcommittee had met the previous week to discuss policy issues regarding the Chief not accepting recommended findings on contested appeals. The consensus [among subcommittee members] was to not pursue changes to city code or union contracts; that's not something that could be accomplished overnight. Rather, the recommended finding from City Council [acting as PIIAC] is that he personally appear in a Council session to explain his rationale . . . for now, this appears to be the only practical solution. There are philosophical issues involved; if a disturbing pattern develops, this could be a catalyst for change.

Some police and sheriff's departments have attempted to accommodate board members' concerns.

- When the Tucson board complained that the police department's data were difficult to follow, the chief had the board chairperson meet with the department's statistical department, which developed a clearer way of presenting the information for the board. The department also offered to provide a photographer for a board project to develop a video for schools designed to improve police relations with the Hispanic community.

- The San Francisco Police Commission required the department to issue a general order mandating that officers submit to an Office of Citizen Complaints interview because officers had been rendering OCC impotent by not showing up. The department now responds to an officer's first violation of an order to appear with an admonishment, the second with a reprimand, and the third with a 1-day suspension. When some officers who did not care about losing a day's pay continued to ignore the summons, the chief told them violations would influence their promotion opportunities.

- Some departments also make special efforts to ensure their officers are familiar with the oversight body's responsibilities and the officers' obligations to it.

   — At the San Francisco Police Department's invitation, Mary Dunlap teaches a 50-minute session at the police academy on "How to Avoid OCC."

   — The Minneapolis Police Department's training commander arranged for Patricia Hughes, the Civilian Police Review Authority's (CRA's) executive director, to give a 90-minute presentation at

> *When some officers who did not care about losing a day's pay continued to ignore the summons, the chief told them violations would influence their promotion opportunities.*

each police academy to explain to recruits how to stay out of trouble and what the citizen review process is. Hughes presents eight sustained cases that have come before CRA, asking the recruits what they would do in each situation. She compares their responses with what the police officers actually did in the situations. An example of a case Hughes presents is described below.

After several squad cars responded to a gang shooting, the last officer to arrive saw a male in the back of a cruiser; the officer pointed his flashlight at the man, and the man made an obscene gesture to the officer. "What would you do?" Hughes asks. After getting the recruits' responses, Hughes continues. "The officer walked around the car, opened the door, and punched him. Later, he learned that the other officers had stowed the man in the cruiser because the person had witnessed the shooting and was terrified he would be retaliated against if the gang members knew he was going to testify. So, the man overreacted when the officer made him visible with his flashlight." Hughes then explains how citizens feel when officers engage in misconduct.

— According to Orange County Deputy Patrick Reilly, "All deputies know the [Citizen Review] Board [CRB] exists—it is discussed in the academy and during supervisor tests." In addition, CRB is the subject of an entire general order dated August 7, 1997, that begins, "The purpose of this policy is to ensure all agency employees are aware of the Orange County Citizen Review Board (CRB)." After describing the board's composition, the bulletin goes on to observe, "All agency employees shall appear before the CRB when formally notified in writing. Failure to appear may result in disciplinary action."

Other departments have gone out of their way to show good faith in working with their oversight systems.

- Lt. Robert Skomra, former commander of the Minneapolis Police Department's IA unit, on his own initiative, went to every Civilian Police Review Authority meeting, bringing a different IA investigator with him each time. "You can't tell people [i.e., CRA] you're a valuable asset unless you go in person."

- Robert Duffy, when he became chief of the Rochester Police Department, came to a meeting in which his IA investigators were training Civilian Review Board members to introduce himself and to explain how valuable the board's work was to the department.

Working With the Union

The local officers' union can be more important than the chief or sheriff—and have different concerns than the chief or sheriff—in making sure citizen oversight functions properly. The union can challenge the process in the courts, influence line officers to cooperate with or hamper the procedure, and expedite or delay proceedings when it represents subject officers during interviews and hearings. It is therefore crucial for oversight planners, staff, and volunteers to address union concerns about the review process.

Historical conflict between most police unions and citizen oversight bodies

Most police unions have traditionally opposed citizen oversight—often successfully—through litigation or lobbying. In one jurisdiction, the union asked city council members to vote against renewing the appointment of a board member who had had the highest rate of recommendations for sustaining complaints.

However, police unions have increasingly been unable to defeat citizen review proposals. In some jurisdictions, unions have either chosen not to oppose oversight proposals or even supported them. Opposition has declined in part because leaders have decided that a review system was inevitable and because, as one union president said, "We're not getting gored by it." According to a union treasurer in another city, "If there had been no sworn [officers] on the board, the union would have opposed it. But we knew we could not litigate it—oversight was inevitable—so we wanted to make the best of what could have been a bad situation." (See "Not All Police Unions Have Opposed Citizen Oversight.")

Approaches to collaboration

There are several ways in which oversight planners can try to work with the local officers' union.

Involve the union in planning the review procedure

In some jurisdictions, oversight planners have invited union leaders to help design the new oversight system. As a result, although union leaders may not agree with the procedure that is ultimately adopted, they have an opportunity to shape its design, express their concerns, and get to know some of the individuals who may be administering it.

- When planning the Boise oversight procedure, Pierce Murphy, the new ombudsman, invited the union president to accompany him to examine the Civilian Police Review Authority in Minneapolis. The Boise Police Department agreed to pay for the president's time and to reimburse his travel expenses.

- In 1997, the Albuquerque City Council established an ad hoc committee on public safety consisting of three city counselors and staff to develop a citizen oversight procedure. The committee in turn assembled a task force of seven individuals representing community organizations (e.g., the American Civil Liberties Union), the police department, and the officers' union. The group met every 2 or 3 weeks for 6 months and ended up presenting five different models to the city council for consideration. A legislative analyst merged the models into a single ordinance, which the council approved.

When possible, accommodate union concerns

Union leaders have legitimate interests in how citizen oversight operates, including:

- The use of subpoena power.

- The system's authority to impose discipline.

NOT ALL POLICE UNIONS HAVE OPPOSED CITIZEN OVERSIGHT

Union opposition to citizen review has never been monolithic. In particular, some police unions representing minority officers have supported citizen review as a means of reducing alleged police misconduct toward racial and ethnic minorities.

- According to the National Black Police Association, an advocacy organization composed of 150 chapters representing more than 30,000 African-Americans in law enforcement, there is "substantial evidence that the police department and its leadership cannot properly discipline their colleagues." The organization executive director goes on to report, "Most traditional police associations and police unions are strongly opposed to citizen's review of police . . . [but] the National Black Police Association . . . strongly support[s] the implementation and use of civilian review of police misconduct."

- A July 11, 1999, *New York Times* article reported, "Lieutenant Eric Adams, president of a civic group representing New York's black police officers, said . . . [a United States Attorney's] investigation [into the New York Police Department's handling of brutality complaints] gives credence to complaints long voiced in minority communities. He said the city needed an independent agency with power to gather evidence, because as long as investigators depend on the Police Department for information, their work will be compromised."

- The African-American Officers for Justice, composed of San Francisco Police Department black officers, joined with liberal organizations in the city in 1982 to urge the Board of Supervisors to place citizen oversight on the ballot as a voter initiative.

Not all rank-and-file members support their union's opposition to citizen review. A sergeant in one jurisdiction reported he did not take a union representative with him to a board hearing on a citizen's complaint because he disagreed with the union's "rightwing positions."

- Delays in case processing.

- The fairness of the procedures.

For example, according to Liana Perez, Tucson's auditor, "It's better not to try to get subpoena power because then the union ties you up in court." Furthermore, many oversight staff and researchers believe that subpoena power is not a particularly useful tool. (See the discussion of subpoena power in chapter 5, "Addressing Important Issues in Citizen Oversight.") Other examples of meeting the union half way follow.

- The Rochester corporate counsel worked closely with the city council and the mayor to fashion an oversight procedure that would not be subject to a successful union suit. For example, the city council attorney reported that giving the Civilian Review Board authority to establish mandatory findings and discipline would make it vulnerable to litigation by the union, and the union already had said it would sue if the council granted the board these powers. As a result, this authority was omitted from the legislation.

- Rochester's first review board included police officers to accommodate union concerns. Later, the city council enacted legislation excluding them. The Rochester police union consented to a mediation option after the city council agreed to exclude any final signed agreement and to call the process "conciliation," not "mediation."

- Board members in another jurisdiction agreed to base their findings on clear and convincing evidence rather than on a preponderance of the evidence after the union agreed that officers could be required to testify.

- To address the concerns that the oversight system could cost the union money, the Minneapolis statute requires that the Civilian Police Review Board pay the union's legal fees when a complaint is not sustained after the executive director has found probable cause.

- When first appointed, Lisa Botsko, Portland's first auditor, met with the union president and treasurer initially "just to get to know them, learn their perspectives, and tell them what I do." One of the grievances they expressed to Botsko was how long it took to process complaints. As a result, Botsko developed timelines for everyone involved in the review process.

Make sure the review process is scrupulously fair

Some union leaders report they have no objection to citizen review as long as it treats officers fairly.

- I wasn't opposed to citizen review; I had researched it and hadn't found that citizen review burns cops. I just didn't want it to be one sided.
  —Butch Swinteck, treasurer, St. Paul Police Officers' Union

- The concept of oversight is excellent; it builds citizen trust in the department: We can't be accused of covering anything up. The problem lies in the practice—the procedure attracts staff with an axe to grind against us. For civilian review of law enforcement to function properly, all parties to it, including officers, must perceive the process to be impartial and professional.
  —John Evans, officer representative, San Francisco Police Officers' Association

According to Mark Gissiner, president of the International Association for Civilian Oversight of Law Enforcement from 1995–99, "Police officers feel that oversight has a third standard for judging officers in addition to criminal behavior and violation of general orders. But there should not be any other standard; either the behavior is a crime or a violation of general orders—or it is not. A lot of people feel oversight should be an advocate for complainants because no one else seems to represent them. But oversight should not; it should be a factfinder only with the authority to make recommendations based on its findings." Of course, once an oversight body finds for the complainant, depending on its legislative mandate, it may have to end up acting on the complainant's behalf.

According to Lt. Steve Young, vice president of the Grand Lodge, Fraternal Order of Police, in addition to basic fairness, "Union leaders are concerned that any oversight system follow due process in how they treat officers." Oversight bodies should, in consultation with the union, investigate thoroughly the due process rights of officers reflected in State statute, case law, and labor's contract with the department. For example, some union contracts require that matters of discipline must be dealt with during working hours. Other contracts limit the information oversight bodies can disseminate to complainants or the general public. The oversight director can distribute a summary of these rights to all staff and volunteers, ask the union to report when it believes a right has been violated, and monitor how well staff are respecting these rights.

Oversight planners can address union concerns about fairness and due process in part by working out a mutually satisfactory arrangement for officer representation at investigatory interviews and hearings that, at a minimum, respects the union contract's requirements regarding officer representation in disciplinary investigations and hearings. The extent to which union leaders may accompany and represent officers at oversight investigatory interviews and hearings varies considerably. In Berkeley, subject officers may have representation by a union agent or the union's legal counsel during all investigations and hearings. Union representatives may attend interviews in Minneapolis, but they may not speak during the investigation (they may caucus to offer the officer advice).

Fairness may always be compromised if political leaders and local officials see oversight simply as a means of placating citizen groups that are asking for a review system. Steve Young observes, "Most union people feel officers are the victim of politics—that city and county officials implement oversight to pander to the complaints of a few vocal citizens and citizen groups. As a result, officials need to articulate nonpolitical, legitimate reasons for implementing it."

Highlight shared objectives

Unions and oversight bodies share the same concern that internal affairs treat officers fairly. As a result, some union leaders have used the citizen oversight system to seek redress for their members whom they felt IA treated unfairly.

- In Portland, the union treasurer—over the objections of some rank-and-file union members—filed a complaint against a lieutenant with the oversight body after internal affairs rejected the complaint without investigation on the grounds that it had no merit. The oversight body voted to send the complaint to IA for investigation.

- When the union president told the Portland auditor about a case he felt IA had handled poorly, Lisa Botsko asked for the case, saying, "That's my job, and I'll audit it." She did, and reported, "The union was right about how the case was mishandled."

- Although he opposed—and still objects to—having an auditor in Tucson, Mike Gurr, the officers' association vice president, has since asked the auditor, Liana Perez, to review sustained cases that he felt IA had not handled well, even though the city manager and police department objected to his referring the cases.

When police and sheriff's departments already are doing a good job investigating citizen complaints, another objective that oversight bodies and union leaders share is to reassure a frequently skeptical or hostile public, or certain community groups, that the department is in fact doing a good job. According to Thomas Mack, treasurer of the police federation that represents Portland Police Bureau officers, "The auditor's review of investigations is good because it opens up the files so people can see the department isn't covering up. And with community policing, it makes sense to look at what IA did. Many [union] members believe IA's review is enough, but I feel, 'Let them take a look.'"

Time may help but is not the cure-all

Some unions may modify their views about the oversight procedure after they have had a chance to see it in operation and find that its staff are unbiased and competent. At the union's insistence, the St. Paul City Council agreed to include a 1-year sunset clause in its oversight legislation. At the end of the year, union leaders realized that the review process tended to be more lenient with officers than internal affairs and decided not to oppose its being made permanent.

Despite everyone's best efforts, some areas of disagreement are likely to remain between oversight planners and staff on the one hand and union leaders and members on the other hand. Even when union leaders themselves may not regard oversight as a cause for serious concern, members may still be leery. According to Capt. David Friend, president of the local police union in Omaha:

> Oversight has a necessary function and its role is important, but my membership is not thrilled because the board is mostly civilians. The CCRB [the Omaha oversight body] finds for the officer in 99 percent of cases, but it's hard to convince cops the board isn't out to get them. Their feeling is, "Just because you're paranoid doesn't mean they're not after you." So they still have a fear that they could be the first to get skewered.

Notes

1. "A major problem with civilian review from the police perspective is the lack of a working understanding of the environment in which [police] decisions are made." "External Police Review: A Discussion of Existing City of Tucson Procedures and Alternative Models, Report to the Mayor and Council," 1996: 12.

2. For example, "The [Massachusetts] Board of Registration in Medicine, . . . according to a growing number of critics, was lax to the point of negligence in policing doctors," *Boston Sunday Globe*, March 28, 1999: B1.

3. The argument that there should be citizen review of police departments because officers alone have the legal authority to exercise lethal force is also specious: doctors, airline pilots—even taxicab drivers—are also in positions that place the lives of their clients in jeopardy.

4. "External Police Review," 12.

5. "External Police Review," 15.

6. Some officers report that, because they are afraid that oversight findings will accumulate in their personnel files and hamper promotion or transfer to desired details, they enforce the law less vigorously. However, "There is no evidence gleaned from civilian review studies that would suggest that this piece of subcultural wisdom is anything but folly . . . the only police review systems that have generated any significant amount of counterproductivity have been internal systems." (Perez, Douglas, *Common Sense about Police Review*, Philadelphia: Temple University Press, 1994: 161–162.) According to Prentice Sanders, San Francisco assistant chief, "Regarding officers who say OCC ties their hands and they cannot do their jobs, I ask: 'How come there are many officers with lots of arrests and no complaints?'"

Chapter 7: Monitoring, Evaluation, and Funding

KEY POINTS

- To justify their funding, oversight bodies need to be able to document their effectiveness. To do so they need to collect monitoring data.

- Efforts to monitor oversight bodies should address:

  — The simplicity, speed, and courteousness of the intake process.

  — The quality of investigators' work product.

  — The performance of board members.

- To evaluate the effectiveness of an oversight system, it first is necessary to establish the objectives the procedure is designed to achieve—something few oversight planners have done.

- Comprehensive evaluations of citizen review have been rare. However, an evaluation of Albuquerque's oversight mechanisms is a good example of a thorough assessment.

- Jurisdictions can implement inexpensive "customer satisfaction" surveys that can suggest how the oversight process may be improved.

- Jurisdictions need to fund their oversight procedures adequately to make sure the mechanisms are effective and respected.

- Funding for oversight ranges from $20,000 (e.g., because the effort is run almost entirely by volunteers) to more than $2 million. Determining funding levels depends on the activities the oversight system will undertake and several other considerations.

- There is an inconsistent relationship between the type of oversight system and cost, although costs are generally higher when oversight involves investigating citizen complaints.

- It is difficult to predict oversight costs before determining what the system's features and activities will be.

- Although more money may not buy more oversight activity, an underfunded procedure may be doomed to failure—and may create more controversy around police accountability than it resolves.

Some programs can flourish without any evidence of effectiveness simply because they are seen as politically useful. Certainly, oversight systems can fall into this category. However, most of the time securing adequate funding for a program depends at least in part on being able to document that it is achieving its objectives. Documenting success in turn can be done only if program activities are closely monitored. This chapter suggests methods of monitoring and evaluating citizen oversight as a prelude to discussing funding issues.

Monitoring

Examining several aspects of a citizen oversight system can suggest how well the process is operating.

The intake process

Intake staff can discourage would-be complainants by an indifferent attitude, lack of helpfulness, or dilatoriness. Program supervisors can assess intake staff performance by casually or formally observing the process. They also can require investigators and board members to ask complainants whether they found the intake process complicated or discouraging. Customer satisfaction surveys (see next section) can include questions about the intake process. Oversight directors can also use "testers"—fake complainants—to monitor the intake process (see "Using 'Testers' to Monitor the Intake Process").

Oversight bodies to which internal affairs units refer complainants can monitor how conscientiously IA investigators are making referrals. The program coordinator of Rochester's Civilian Review Board (CRB) asks each chairperson to visit the police department's IA section four times a year to review whether a random sample of six cases that were never sent to CRB (e.g., because a citizen dropped the complaint) should have been forwarded. CRB administrators talked with IA on the one occasion in which a chairperson felt a case should have been referred because of its sensitive nature.

Investigators' work

Most programs have a procedure for reviewing the quality of their investigators' work. Many tape record all interviews, not only to have a permanent record of what complainants, officers, and witnesses said but also to review the investigators' techniques.

In Flint, the senior investigator reviews the findings of every investigator, and the ombudsman provides a final review. In San Francisco, one of three Office of Citizen Complaints (OCC) senior investigators reviews every file, followed by a review by the chief investigator. If investigators recommend the complaint be sustained, one of two OCC attorneys reviews the case to assess whether the evidence is sufficient and the findings comply with applicable laws, rules, and orders. Mary Dunlap, the OCC director, reviews the file for a final determination. Dunlap reviews about 1,500 packets a year, averaging 6 per working day. It takes her 3–5 minutes to review simple cases, but complex and important cases can take many hours. Because she feels that supervision of investigators is critically important for quality control, Dunlap makes sure she hires enough supervisors and provides them with extensive training and close oversight. Each senior investigator is responsible for the same five investigators' work product so they can monitor the investigators' progress over several months or longer.

Any of the OCC reviewers may send inadequate packets back to investigators for additional work. The most typical—although still uncommon—problem supervisors

> *In San Francisco, one of three Office of Citizen Complaints (OCC) senior investigators reviews every file, followed by a review by the chief investigator. Mary Dunlap, the OCC director, reviews the file for a final determination.*

USING "TESTERS" TO MONITOR THE INTAKE PROCESS

To monitor the intake process, program directors can arrange for unknown citizens to file fictitious complaints. Although using testers may anger program staff, the technique is common in the private sector when retail businesses need to make sure their personnel are providing good customer service. Testers can provide valuable information regarding the thoroughness with which staff screen citizens' complaints, how well they help complainants fill out necessary forms, how accurately and politely they answer questions, and how expeditiously they meet complainants' needs.

find is that an investigator failed to ask the complainant or subject officer an important question.

Board members' performance

Oversight directors typically attend all board hearings, giving them an opportunity to observe board members' behavior. Directors need to be especially sensitive to whether members appear biased for or against the police (see "Can Repeated Contact With Police Officers Impair Oversight Staff Objectivity?").

In Rochester, board members meet privately without the director's presence. As a result, chairpersons are required to report another board person who appears to be biased to the program coordinator, who then will meet with the person to discuss the matter. A few board members have been dismissed when it was discovered they discussed a case with a third party. In another jurisdiction, the director removed a board member who had told a police officer that she did not like his behavior, warning him, "And I'm a member of the citizen oversight board."

Consumer Satisfaction Surveys

Jurisdictions can inexpensively implement a customer satisfaction survey. The Vera Institute of Justice in New York City surveyed a sample of 371 citizens who had filed complaints with the Citizen Complaints Review Board. The study found:

- Most complainants (61 percent) had "moderate" objectives: an apology for themselves or a reprimand of the officer(s).

- The desire for a direct encounter with the subject officer(s) was "pervasive" and "significantly associated with complainant satisfaction."[1]

Minneapolis' Civilian Police Review Authority (CRA) hired Samuel Walker, a consultant, to develop two customer satisfaction surveys. The board and union attorney reviewed drafts of the surveys. Patricia Hughes, the executive director, sends one survey twice a month to the previous week's complainants and the second survey (see exhibit 7–1) to complainants and officers after their cases have been settled. The anonymous survey includes an addressed, stamped envelope to be mailed back to the city coordinator's office, which tabulates the responses and sends a copy to Hughes.

> *Minneapolis' Civilian Police Review Authority hired a consultant to develop two customer satisfaction surveys.*

Walker's analysis of 29 surveys completed by citizens and 21 completed by officers found:

- Although most citizens were satisfied with how CRA staff treated them, 8 of the 29 respondents reported they were not treated with respect.

CAN REPEATED CONTACT WITH POLICE OFFICERS IMPAIR OVERSIGHT STAFF OBJECTIVITY?

A close observer of citizen oversight has reported that investigators' and board members' "daily interactions with the police force allow them plenty of opportunities to develop empathy and subliminal ties with those involved in 'real law enforcement.'"* For the same reason, some citizens have objected to oversight staff attending police citizen academies (see chapter 4, "Staffing"). A report proposing the redesign of Minneapolis' Civilian Police Review Authority (CRA) recommended that "Periodic monitoring should be done to ensure that co-option does not become an issue (regarding staff)." However, CRA responded, "There exists no known, objective, scientific measure for 'co-optation.' Therefore, 'monitoring' of this possibility must be subjectively assessed by the Executive Director of the CRA and its Board members."

---

* Perez, Douglas W., *Common Sense About Police Review*, Philadelphia: Temple University Press, 1994: 182–183.

Chapter 7: Monitoring, Evaluation, and Funding

Exhibit 7–1. Minneapolis Consumer Satisfaction Post-Outcome Survey

DRAFT

1. Do you feel you had a chance to tell your side of the story? Yes ___ No ___

2. Do you feel you were treated with respect? Yes ___ No ___

3. If you accepted <u>mediation</u> of your complaint, was the mediation successful? Yes ___ No ___

4. If your complaint resulted in a <u>hearing</u>, were you satisfied with the hearing process? Yes ___ No ___

5. Do you feel the outcome of your contact with the CRA was <u>fair</u>? Yes ___ No ___

6. Is there anything you would like to tell us about your experience with the CRA?

   _____
   _____
   _____
   _____
   _____

For our records, we would like to know a few things about the nature of your complaint.

7. My complaint involved

   ___ Excessive Force   ___ Inappropriate Conduct   ___ Harassment
   ___ Inappropriate Use of Force ___ Failure to Provide Service ___ Theft
   ___ Inappropriate Language   ___ Discrimination

8. I am: Male ___ Female ___

   African American ___ Asian American ___ Hispanic/Latino ___ Native American ___ White ___

   Under age 18 ___   18-24 ___   25-34 ___   35 or older ___

Please return to the City Coordinator's Office, 350 South Fifth Street, Room 301M, Minneapolis, MN 55415, in the enclosed business reply envelope.

Thank you!

- Almost half of the citizen respondents felt the outcomes of their cases were not fair.

- Eighteen of the 21 responding officers felt the outcomes of their cases were fair.

Because only 29 of 174 citizens and 21 of 81 police officers to whom the survey was sent returned them, it is difficult to know whether the responses are representative of more general satisfaction or dissatisfaction among all complainants and officers. Nevertheless, the responses are useful for documenting considerable satisfaction with CRA while pointing to areas for possible improvement. Walker plans to reexamine the data after a year when 140–150 surveys should be available for analysis, at which time he will examine levels of satisfaction in relation to the citizens' and officers' gender, race, and age.[2]

A Tucson city councilman telephones each constituent in his district who appeals an IA investigation to see if the person was satisfied with the auditor's review. For example, he called a mother who complained about an officer pulling over her son. The woman reported she had learned from the auditor that her son had been in possession of an illegal substance and had no valid driver's license, facts of which she had not been aware. As a result, she now was satisfied with the officer's behavior—and pleased with the auditor's review. Every constituent but one whom the councilman has telephoned has reported being satisfied with the oversight process.

Evaluating the Citizen Oversight Process

Two long-time observers of citizen review have reported, "There are no thorough, independent evaluations of the effectiveness of any [citizen oversight] procedures, much less any comparative studies."[3] Although one comprehensive evaluation has been conducted since this statement was made, the lack of evaluations is perhaps not surprising.

Establishing objectives

To evaluate the effectiveness of an oversight system, it first is necessary to establish the objectives the procedure is designed to achieve—something few oversight planners have done. The objectives need to be specific and measurable. However, this crucial step is frequently omitted, making it necessary to develop objectives for measuring program success after the fact. In addition to hampering any evaluation, not establishing objectives from the outset leaves program staff uncertain—or mistaken—about what they are supposed to be doing.

> *To evaluate the effectiveness of an oversight system, it first is necessary to establish the objectives the procedure is designed to achieve—something few oversight planners have done.*

Among the possible objectives planners can establish for citizen review of the police are the following:

1. Increase the public's confidence that the police or sheriff's department is addressing citizen complaints fairly and thoroughly.

2. Reassure the public that the police or sheriff's department appropriately disciplines officers who engage in misconduct.

3. Defuse hostility toward public officials or the police or sheriff's department expressed by residents or specific groups of residents. This typically is the reason most oversight systems are established.

4. Improve the fairness and thoroughness of the police or sheriff's department's investigations of citizen complaints (for example, by auditing the department's own procedures or conducting investigations in tandem with or instead of internal affairs department investigations).

5. Reduce misconduct by police officers, such as verbal abuse, use of excessive force, and discriminatory enforcement of the law.

6. Reduce the number of police shootings.

7. Help ensure that officers who engage in misconduct are appropriately disciplined.

8. Provide the public with an understanding of the behavior of police officers and sheriff's deputies.

9. Provide the public with a "window" on how the police or sheriff's department investigates allegations of officer misconduct.

10. Increase legitimate citizen complaints by, for example:

    - Providing an avenue for filing complaints that is less intimidating than going to the police or sheriff's department.

    - Increasing confidence that complaints will be taken seriously.

    - Providing more accessible locations for filing.

11. Provide an open and independent forum for the public to express general concerns about the police or sheriff's department's operations or about officer conduct.

12. Provide a mechanism through which citizens can suggest recommendations for improving police policies and procedures and police training.

13. Establish a mediation option for resolving selected complaints to achieve one or more of the potential benefits of mediation identified in chapter 3.

After establishing objectives, program planners and administrators need to determine how they will know whether each one has been achieved in terms of:

- What level of activity will be considered a success (e.g., 10 fewer police shootings in each of the next 2 years compared with the average number of shootings during each of the previous 10 years; a 15-percent increase within 2 years of citizens who feel confident the department is disciplining officers appropriately).

- How the necessary data will be gathered (e.g., police records of use of firearms over a 12-year period; public survey of community attitudes toward the police before the oversight procedure was initiated and 2 years later).

There are other barriers to evaluation in addition to the failure to develop measurable objectives. Many administrators consider monitoring and evaluating program activities a low priority. They may lack the time, money, or expertise to assess their programs, or they may be concerned that negative results may jeopardize their funding and even their positions. Finally, even with solid objectives and the time, skills, and will to conduct a useful evaluation, there may be problems with the data that make it difficult to draw valid conclusions about the oversight system's effectiveness (see "Data Barriers to Evaluating Oversight Procedures").

> The Albuquerque City Council commissioned a $27,602 evaluation of its oversight system and used many of the findings in restructuring its procedures.

The Albuquerque evaluation

Despite these concerns and barriers, assessing program effectiveness is essential for learning how to improve oversight operations and for demonstrating that the oversight process should be maintained or expanded. Reflecting this need, the Albuquerque City Council commissioned a $27,602 evaluation of its oversight system and used many of the findings in restructuring its procedures (see "The Albuquerque City Council Commissioned a Thorough Evaluation"). Although this approach may cost more money than jurisdictions are able to spend, other cities and counties easily can afford to implement parts of the evaluation.

Funding

Oversight bodies have dramatically different budgets: Some, such as the review board in Orange County, cost relatively little because they rely almost entirely on volunteers and in-kind services, while others, such as those in Flint and Minneapolis, run into the hundreds of thousands of dollars (see exhibit 7–2). San Francisco's oversight budget is more than $2 million.

The relationship between oversight costs and oversight activity

In most organizations, there is a relationship between expenditures and results—that is, the more money spent, the more or better the results. Are increased expenditures for oversight associated with increased *utilization*—that is, do oversight systems get what they pay for? Exhibit 7–2 presents the nine oversight systems

DATA BARRIERS TO EVALUATING OVERSIGHT PROCEDURES

It would seem logical to evaluate the success of citizen oversight by determining whether complaints increase or decrease after the system becomes operational. The natural assumption is that, over time, complaints will decline as the oversight system begins to play a part in reducing officer misconduct. However, complaints may increase because the intake process has been simplified or made more accessible or because public confidence in the review process has increased. Furthermore, there are rarely complete—or any—baseline data on the actual number of incidents of police misconduct.*

Case review data also are suspect. The number of cases reviewed by the Rochester Civilian Review Board declined significantly one year and rose the next. The decline occurred because there had been several shootings, and IA investigators had to drop less serious investigations and delay examining new ones while they investigated the complex and high-profile shootings. Staff turnover among investigators further delayed investigations. More cases were investigated the following year after staffing problems ended and investigators caught up on their backlog of less serious cases.

Examining changes in community attitudes toward the police would provide more valid information on oversight effectiveness. However, these data, too, may be misleading because many other events may be taking place simultaneously in the community that could change the public's attitudes toward the police, including community policing or a new police chief who is more strict or more lax about discipline than his or her predecessor. Finally, it is difficult to compare the effectiveness of different citizen review models because they have different goals, resources, and constraints.

---

* Walker, Samuel, and Vic W. Bumphus, "The Effectiveness of Civilian Review: Observations on Recent Trends and New Issues Regarding the Civilian Review of Police," *American Journal of Police* 11 (4) (1992): 1–26.

---

arranged in ascending order of budget levels along with their activity levels for 1997. As shown, it is impossible to compare activity levels among oversight systems because different systems engage in different types of activities—for example, investigations, hearings, mediations, and audits. Nevertheless, it is possible to examine five different relationships between oversight budgets and activity levels.

1. There is no association between budget levels and overall activity levels among the nine systems. For example, Rochester's level of activity (26 cases reviewed, 4 cases mediated) appears to be less than Orange County's, Portland's, and St. Paul's, even though its budget is much higher. Tucson, with a budget of more than $144,000, monitored 63 investigations, while Portland, relying on a single staff person's salary of $43,000, audited 98 cases and processed 112 appeals.

2. There is a clear association between higher budgets and whether a system conducts investigations: All four systems with the highest budgets are type 1 systems (which investigate citizen complaints). These systems are the most expensive because they have to hire professional investigators rather than rely on volunteers. It also appears that, the higher the 1997 budget among these four type 1 oversight systems that conduct investigations, the more overall activity (not just investigations) they engaged in. For example, while Flint conducted more investigations (313) than either Berkeley (42) or Minneapolis (159), Berkeley conducted 12 hearings and 34 preliminary investigations, and Minneapolis provided assistance to 715 citizens as well as arranged for 14 cases to be mediated. (If the Minneapolis and San Francisco oversight bodies did not conduct investigations, their respective police departments would have to hire additional internal affairs investigators [unless they could transfer existing personnel] to perform the work the oversight bodies had been conducting. As a result, the additional money that oversight bodies need to conduct investigations does not cost the city, town, or

THE ALBUQUERQUE CITY COUNCIL COMMISSIONED A THOROUGH EVALUATION

Troubled by fatal shootings by Albuquerque police officers (31 in 10 years), extremely high annual payments for tort claims involving police officers (up to $2.5 million per year), and other concerns, the Albuquerque City Council in 1996 hired Eileen Luna and Samuel Walker, two well-known experts in citizen review of police, to evaluate the city's oversight mechanisms. Luna and Walker's 159-page report concluded: "The existing mechanisms for oversight ... are *not functioning effectively*" (emphasis in original).[1]

The researchers used five sources of data:

1. A survey administered in person to more than 357 rank-and-file officers (44 percent of the total sworn officer force) consisting of 70 close-ended and several open-ended questions.

2. Personal interviews with:

   - Community members, including leaders and members of human rights, civil rights, and neighborhood organizations; attorneys in private and public practice; and spokespersons for ethnic communities.

   - Police officers, including the chief, command officers, union officers, and the staff psychologist.

   - Public officials, including the mayor, city council members, and advisory board members.

3. A review of official documents, ranging from advisory board minutes to internal affairs quarterly reports.

4. An audit of the internal affairs unit's general patterns and practices, including:

   - Complaint files for 1994–96.

   - A consumer satisfaction survey administrated to everyone who had filed a complaint during the previous 3 years.

5. A national survey of citizen oversight mechanisms.

The report concluded with 10 recommendations for improving the oversight process in Albuquerque, including advocating that oversight procedure administrators exercise fully the authority they already had and that the mayor and city council take a more active role in overseeing the police department.

In October 1998, the city council enacted the Police Oversight Ordinance to restructure the city's oversight system along the lines of the report's recommendations. According to coauthor Samuel Walker, "The City Council is to be commended. They paid an outsider to come in and then acted on those [recommended] changes. Very often these reports just sit on a shelf."[2] The council provided for an 18-month evaluation of the new oversight system.

---

1. Luna, Eileen, and Samuel Walker, *A Report on the Oversight Mechanisms of the Albuquerque Police Department*, prepared for the Albuquerque City Council, 1997.

2. *Law Enforcement News*, "Who's Watching the Watchers?" 29 (499) (November 15, 1998).

EXHIBIT 7–2. COSTS OF NINE OVERSIGHT SYSTEMS IN 1997 IN RELATION TO RESPONSIBILITIES AND ACTIVITY

Location	Name	Cost in 1997	Principal Responsibilities and Features	Activity in 1997	Mean Cost per Complaint Filed or Reviewed
Orange County pop: 749,631 sworn: 1,134	Citizen Review Board	$20,000	• board members review IA findings on use-of-force cases • board members make policy recommendations • a sheriff's captain coordinates the board's activities	45 board hearings } 45	$444
St. Paul pop: 259,606 sworn: 581	Police Civilian Internal Affairs Review Commission	$37,160[a]	• police chief established board as in-house review procedure • board, not IA, recommends discipline • seven-member board includes two St. Paul police officers	71 cases reviewed } 71	$523
Portland pop: 480,824 sworn: 1,004	Police Internal Investigations Auditing Committee	$43,000	• the city council hears appeals • citizen advisers review closed cases • an examiner audits IA investigations • all three provide policy recommendations	21 appeals processed 98 audits of completed cases } 119	$361
Rochester pop: 221,594 sworn: 685	Civilian Review Board	$128,069	• board members review IA findings on use-of-force cases • board members are trained mediators • specially trained board members conduct conciliations • board recommends policy and procedure changes	26 cases reviewed 4 cases mediated } 30	$4,269
Tucson pop: 449,002 sworn: 865	Independent Police Auditor and Citizen Police Advisory Review Board	$144,150	• auditor reviews completed cases • auditor sits in on ongoing cases and asks questions • board hears community's concerns about the police • board can review completed cases • auditor and board make policy recommendations	289 citizen contacts (9/1/97 to 6/30/98) 96 complaints filed 63 investigations monitored } 159	$755[b]
Flint pop: 134,881 sworn: 333	Office of the Ombudsman	$173,811[c]	• office investigates complaints against all city departments • office reports subject officers' names to the media • chief relies primarily on internal affairs' (IA's) investigation results	313 cases against police investigated (1996) } 313	$555

CHAPTER 7: MONITORING, EVALUATION, AND FUNDING

EXHIBIT 7-2. COSTS OF NINE OVERSIGHT SYSTEMS IN 1997 IN RELATION TO RESPONSIBILITIES AND ACTIVITY (CONTINUED)

Location	Name	Cost in 1997	Principal Responsibilities and Features	Activity in 1997	Mean Cost per Complaint Filed or Reviewed
Berkeley pop: 107,800 sworn: 190	Police Review Commission	$277,255	• board hears complaints in public hearings • board and IA investigate many cases simultaneously • board recommends policy changes	42 complaints filed and investigated 34 cases closed for lack of merit or complainant cooperation }57[d] 12 hearings	$4,864[e]
Minneapolis pop: 358,785 sworn: 919	Civilian Police Review Authority	$504,213	• paid professionals, not IA, investigate complaints • volunteer board members hear complaints with probable cause • half of cases with probable cause are stipulated • many cases are professionally mediated	715 citizen contacts 159 formal complaints investigated }159 14 cases mediated 6 stipulations 3 hearings	$3,171[f]
San Francisco pop: 735,315 sworn: 2,100	Office of Citizen Complaints	$2,198,778	• office, not IA, investigates most complaints • office prosecutes cases in chief's hearings and police commission trials • office recommends policy changes	1,126 complaints received 983 cases investigated and closed }983 50 chief's hearings 6 police commission trials	$2,237

a. Because the board has never spent its $10,000 allocation for hiring an independent investigator, its true budget has been $27,160.
b. Based on a 10-month prorated budget of $120,058.
c. This represents the proportion of staff who handle complaints against the police.
d. Because some of the 42 cases filed and investigated were among the 34 cases closed, the total activity does not represent the sum of the 42 complaints investigated and the 34 cases closed.
e. The mean cost per complaint filed in 1997 was unusually high because oversight staff devoted considerable time to labor-intensive activities beyond investigating complaints, such as conducting a major pepper spray study, staffing a medical marijuana task force, and holding a public hearing on a use-of-force case.
f. This figure is misleadingly high as a measure of overall oversight costs because it does not include the 556 other contacts authority staff had with the public that did not result in formal complaints.

county more money. It merely saves the law enforcement agency money.)

3. It is theoretically possible to compare the nine oversight budget levels with the number of units of a specific type of service provided. The most comparable unit of service to examine is the number of complaints filed or reviewed.[4] Exhibit 7–2 shows the units of this type of service for each of the nine oversight systems. As shown, there is no association between budget levels and number of cases filed or reviewed. Portland, whose oversight systems costs $43,000, reviewed 119 cases, while St. Paul, with a budget of $37,160, reviewed 71 and Rochester, with a budget of $128,069, reviewed 26 and mediated 4. Flint handled 313 cases with a prorated budget of $173,811 (the system investigates complaints against other city agencies), while Minneapolis, with a budget of more than $500,000, handled 159.

4. The final column in exhibit 7–2 shows how much it costs to process each complaint that a citizen filed or oversight system staff reviewed in 1997 by dividing the system's budget by the number of filed complaints or complaints reviewed.[5] As shown, there is no correlation between cost per complaint filed or reviewed and overall activity level. For example, Rochester's Civilian Review Board, which reviewed cases in 1997, had a unit cost is $4,269, while the oversight systems in Orange County, Portland, and St. Paul reviewed or heard between 45 and 119 complaints with a cost per complaint filed of less than $525. A similar discrepancy appears for Flint ($555 for each of 313 complaints) and Berkeley ($4,864 for each of 57 complaints).

5. Finally, as shown in exhibit 7–2, there is no relationship between a system's budget and its cost per complaint filed. San Francisco, with the largest budget, has a lower cost per complaint than Berkeley and Minneapolis, which have smaller budgets. Rochester has a much higher cost per complaint reviewed than Tucson and Flint, even though its budget is lower than theirs.

Why is there a weak correlation at best between budget levels and these measures of oversight system activity?

- Oversight systems often cannot alter the number of complaints they investigate or review even with increased funding because they are limited by statute to accepting only certain types of cases (e.g., use of force—Rochester) or are mandated to accept all cases within certain categories (e.g., use of firearms—St. Paul).

- An oversight system's activity level for a given year may reflect considerations that have nothing to do with budget levels. For example, there often are anomalies in the number of complaints filed in a given year. A large and unruly public demonstration may result in numerous complaints, or there may be a decline in the number of cases forwarded by the internal affairs unit because the unit loses staff or gets backed up investigating high-profile shootings. Changes in oversight staffing levels—for example, through resignations—may affect system activity in a given year irrespective of budget levels.

- More money for citizen oversight may fail to result in increased utilization if staff do not make citizens aware of the opportunity to file complaints, if citizens do not trust the system, if the police or sheriff's department refuses to cooperate, or if other barriers to filing complaints are not addressed.

- Cost-per-complaint figures do not take into consideration each system's total responsibilities or the in-kind services it uses. For example, Berkeley's cost per filed complaint includes not only its cost of investigating the complaint but also of holding hearings and making policy recommendations. Minneapolis spends many hours every year helping hundreds of citizens who decide not to file a complaint.

Does more money buy better *quality* service? Unfortunately, most systems do not monitor the quality of their services. Furthermore, it was outside the purview of this publication to examine quality of services. Finally, it was not possible to examine oversight activity in some jurisdictions—for example, hearings in Rochester are not open to the public.

The relationship between cost and oversight type

There is a *theoretical* relationship between the four types of oversight systems (see chapter 1) and cost.

- Type 1 oversight systems, in which citizens investigate allegations and recommend findings (Berkeley, Flint, Minneapolis, San Francisco), are the most expensive because, as previously noted, professional investigators must be hired to conduct the investigations—lay citizens do not have the expertise or the time.

- Type 2 systems, in which citizens review internal affairs findings (e.g., Orange County, Rochester, St. Paul), tend to be inexpensive because volunteers typically conduct the reviews.

- Type 3 systems, in which citizens review complainants' appeals of police findings (Portland), also can be inexpensive because of the use of volunteers.

- Type 4 systems, in which auditors inspect the police or sheriff department's own complaint investigation process (Portland, Tucson), tend to fall in the midlevel price range. On the one hand, like type 1 systems, only a professional has the expertise and time to conduct a proper audit. On the other hand, typically only one person needs to be hired because the auditing process is much less time consuming than conducting investigations of citizen complaints.

In practice, however, there is an inconsistent relationship between oversight type and cost. This is because, when examined closely, many oversight operations are not "pure" examples of a type 1, 2, 3, or 4 system. As a result, the actual cost for a given type of oversight system may be more or less expensive than what a pure type would cost. For example, San Francisco's type 1 system requires extra staff to prosecute cases at chief's hearings and police commission trials. While type 2 systems (e.g., Orange County, St. Paul) generally cost relatively little, Rochester's budget is more than $128,000 in part because paid staff are involved in training and certifying citizen reviewers as mediators and in subcontracting for mediation of selected cases.

There are other reasons why there is frequently no direct relationship between type of oversight system and cost:

- Flint's ombudsman system probably achieves some economies of scale because it investigates complaints from all city agencies, not just the police.

- The actual cost of any oversight system depends on the number of complaints it hears, which in turn hinges on many considerations that are independent of its budget, such as the kinds of complaints its charter permits it to accept or review and how frequently officers in the local department engage in misconduct. For example, some type 1 oversight systems investigate almost all citizen complaints (e.g., Minneapolis, San Francisco), while others investigate only some (e.g., Berkeley, Flint).

- Type 1 and 2 oversight systems (e.g., Minneapolis) can reduce the expense of holding hearings or conducting reviews by diverting some cases to mediation.

- Labor market rates for investigators and executive directors can influence an oversight system's costs, as can the use of inhouse staff versus contracted staff who receive no fringe benefits.

- Different systems provide different stipends to volunteer board members. For example, Berkeley provides $3, while Rochester provides $35 (totaling $15,800 in its 1997 budget).

In sum, it is difficult to predict what an oversight system's actual costs will be before agreeing on what all its components and activities will be and before selecting from among many mechanisms—and combinations of mechanisms—for paying for oversight activities. That said, it is still important to provide a level of funding that will make it possible to investigate, hear, or audit all the cases a program can expect to handle.[6]

> *It is difficult to predict what an oversight system's actual costs will be before agreeing on what all its components and activities will be.*

The importance of adequate funding

Without adequate funding, the oversight process may be just a political statement without any substance—"See what we're doing to hold the police accountable." According to Samuel Walker, an oversight researcher, when boards lack effective investigative powers, necessary funds, and expertise, "We should really think of that as consumer fraud—it is promising an independent review of complaints and not delivering. And that's really fraud."[7] As a result, the Flint city charter stipulates that the ombudsman "shall be granted a budget adequate to allow such a staff as is reasonable and proper for the performance of the duties of said office." Similarly, San Francisco's voters passed an initiative that requires the city to fund one Office of Citizen Complaints investigator for every 150 police officers.

Funding an oversight system with money from the police or sheriff's department's budget, as is done in St. Paul and San Francisco,[8] may hamper efforts to provide adequate funding for two reasons:

1. The more money chiefs allocate to citizen oversight, the less they have for other important functions. In competing with these other functions, oversight may come out the loser.

2. Departments that have an adversarial relationship with the oversight body naturally will be reluctant to provide any more money for it than they have to.

A 1997 editorial, "The Starving Watchdog," in the *San Francisco Examiner* suggested:

> [T]he [county/city] supervisors should ask why appropriations for the watchdog agency [i.e., Office of Citizen Complaints] must be approved by the object of its investigations, the Police Department, and included in the SFPD's proposed budget. . . . It is asking a lot of Police Chief Fred Lau and his planners to cut their own money requests for the benefit of civilians often derided by cops on the beat as second-guessers.[9]

Furthermore, if the oversight process fails, citizens and public officials can blame the department for not providing it with enough money.

Elected and appointed officials, not just police and sheriff's departments, also may keep funding levels unrealistically low.

- Every year some members of the Minneapolis City Council want to abolish the Civilian Police Review Authority because it is expensive. In the 1996 budget process, some council members argued that CRA was a waste of $400,000 because the police department's IA unit should be investigating misconduct.[10]

- Although San Francisco residents voted in 1995 to require one Office of Citizen Complaints investigator for every 150 officers, the Board of Supervisors did not immediately allocate the money to hire additional investigators and supervisors to monitor them.[11] During several months in 1997, staff attrition reduced the number of investigators well below the charter requirement; at many points, there were only 8 investigators (compared with 19 in 1998). According to Mary Dunlap, OCC's director, "It was a battle [to get the money]." Dunlap met and corresponded with supervisors and mayor's aides, and she testified extensively to the Finance Committee of the Board of Supervisors during budget hearings. The OCC staff attorney organized a letter writing campaign to the mayor and supervisors from about a dozen individuals and community groups, including the American Civil Liberties Union and National Association for the Advancement of Colored People.

The tasks of monitoring, evaluating, and securing funds for an oversight procedure may seem intimidating. However, as the following chapter indicates, there are many resources available that can assist oversight planners to address these and other oversight planning tasks.

Notes

1. Vera Institute of Justice, *Processing Complaints Against Police in New York City: The Complainant's Perspective,* New York: Vera Institute of Justice, 1989.

2. For additional guidelines for program evaluation, see Walker, Samuel, *Police Accountability: The Role of Citizen Oversight,* Belmont, California: Wadsworth, forthcoming.

3. Walker, Samuel, and Vic W. Bumphus, "The Effectiveness of Civilian Review: Observations on Recent Trends and New Issues Regarding the Civilian Review of Police," *American Journal of Police* 11 (4) (1992): 1–26.

4. It is possible to calculate mean costs by using another standard besides complaints filed. However, complaints filed appear to be the most universal activity among oversight systems, if audits and reviews of cases are considered within the definition of "filings."

5. The mean cost per complaint filed may not be correct for all systems. In Minneapolis and San Francisco, hearings and mediations for some complaints that are filed are not conducted until the following calendar year, while some complaints that are heard were filed the previous year. As a result, the mean cost was calculated by dividing the budgets for these programs by the number of cases investigated in 1997, exclusive of the number of hearings and mediations held that year.

6. In its 1996 report to the city council examining reconfigured options for its oversight mechanism, Tucson city staff provided startup, first-year, and recurring cost projections for five different types of oversight responsibilities, ranging from an expanded intake function to subpoena power to an independent auditor model (see appendix E).

7. National Public Radio, Morning Edition, July 31, 1997, "Policing the Police."

8. However, the Board of Supervisors in San Francisco, not the police department, establishes the Office of Citizen Complaints' budget.

9. *San Francisco Examiner,* "The Starving Watchdog," June 17, 1997.

10. To avoid funding cuts, some oversight staff may be tempted to accept cases for investigation or review cases that do not merit intake. One city council calculates its oversight body's cost per complaint. Because the number of complaints declined one year, the cost per complaint rose. As a result, some council members felt the board's funding should be reduced. Funding for Berkeley's Police Review Commission has changed significantly and frequently over the years, for example, going from $346,233 in 1994 to $196,732 in 1996 to $277,000 in 1998.

11. Investigators earn from $46,000 to $56,000, senior investigators earn from $50,000 to $62,000, and the chief investigator earns from $54,000 to $71,000. In Minneapolis, Civilian Police Review Authority investigators earn from $38,000 to $51,000.

Chapter 8: Additional Sources of Help

KEY POINTS

- Information for establishing and improving citizen oversight systems is available from:

  — Organizations.

  — Oversight programs.

  — Publications and reports.

  — Individuals with experience in oversight systems.

- Some jurisdictions have engaged in comprehensive research on their own to determine what type of oversight system would be best for their communities. Their research strategies are instructive.

This chapter identifies resources with information about setting up and improving citizen oversight systems. The resources are based on a limited search and therefore are not comprehensive.

Organizations

The International Association for Civilian Oversight of Law Enforcement (IACOLE) is devoted to advancing the cause of citizen oversight of law enforcement. The organization sponsors an annual world conference for oversight practitioners and researchers, publishes a quarterly newsletter *(International Connection)* with position papers and recent developments in the field, provides information to jurisdictions interested in creating citizen oversight agencies, and provides a compendium of oversight agencies and publications. Address: P.O. Box 99431, Cleveland, OH 44199–0431; phone: 513–352–6240; fax: 513–624–8042; e-mail: IACOLE1@fuse.pnet.

The National Association for Civilian Oversight of Law Enforcement (NACOLE) provides educational opportunities and technical assistance to existing and emerging organizations that perform civilian oversight of law enforcement. The organization provides a national forum for information gathering and sharing for these organizations. Established in 1993, NACOLE supports civilian review boards that are under attack by writing letters of support to the local political establishment and community. In addition, the organization will send someone when possible to testify on the behalf of civilian review boards. The association is considering developing a mentor program within NACOLE to make the informal networking that is occurring formal. Address: 9420 Annapolis Road, Suite 302, Lanham, MD 20706; phone: 317–327–3429.

Selected Program Materials

The appendixes to this report contain a number of materials from oversight bodies studied for the publication. In addition, many oversight bodies have developed detailed reports of their procedures for handling citizen complaints, including the following:

- The Berkeley Police Review Commission's "Regulations for Handling Complaints Against Members of the Police Department" provides 16 pages of detailed procedures for intake, investigations, and board reviews.

- Minneapolis' "Civilian Police Review Authority Administrative Rules" provides 28 pages of guidelines for citizen review. The bound booklet addresses the collection and dissemination of data, definitions, standing to file a complaint, intake, grounds for dismissal, mediation, investigations, personal bias or prejudice, rules of evidence, burden of proof, and disposition.

CHAPTER 8: ADDITIONAL SOURCES OF HELP

> ## ALBUQUERQUE, PORTLAND, AND TUCSON DID THEIR OWN RESEARCH
>
> Albuquerque, Portland, and Tucson conducted reviews of oversight procedures in other jurisdictions to learn how best to improve their own oversight systems.
>
> ### Albuquerque
>
> According to Linda Stewart, an aide to the Albuquerque mayor, because of a rash of police shootings in 1997, the city council established an ad hoc committee on public safety consisting of three city counselors and staff. Members visited San Jose and Long Beach and conducted conference calls with other cities. The city council's legislative policy analyst convened a town hall meeting for 300 people to hear their concerns. Finally, the committee appointed and the analyst assembled a task force of seven individuals representing community organizations (e.g., the American Civil Liberties Union and the police department and union).
>
> The group met every 2 or 3 weeks for 6 months to identify areas of agreement and disagreement in terms of what kind of citizen oversight system to establish. The members reviewed "stacks" of ordinances from other cities and also had an evaluation report assessing the current oversight system. The task force presented five different models to the city council for consideration. The legislative analyst merged the best of the models into a single ordinance, which the council approved.
>
> According to Stewart, "The most important part of the process was inviting the activists who were so dissatisfied with the police to sit down and forcing them to explain what they wanted done." The mayor and chief supported the ordinance, and the mayor was getting ready to sign it.
>
> ### Portland
>
> In 1992, the Portland City Council appointed the mayor to chair the Police Internal Investigations Auditing Committee (PIIAC) with the expectation that she would evaluate its operations and recommend improvements. (Members of PIIAC had resigned in protest, alleging the group was ineffective.) The mayor reviewed recent assessments of the PIIAC process, including a self-assessment by the citizen advisers, the auditor's reports, and proposals from community organizations. She also reviewed citizen oversight systems in other jurisdictions and consulted with citizens who had filed complaints with PIIAC. She attended adviser meetings and the city council's public hearing on PIIAC.
>
> As a result of this research, the mayor prepared a report to the city council that included five pages of recommended changes to the PIIAC process to address primarily three identified PIIAC weaknesses: complainants' feelings of intimidation using PIIAC; the perceived failure of the citizen advisers to address policy issues inherent in cases; and advisers' lack of information by which to assess the quality of IA investigations.

Oversight bodies also produce annual reports. Minneapolis' Civilian Police Review Authority, Rochester's Civilian Review Board, San Francisco's Office of Citizen Complaints, and San Jose's Office of the Independent Police Auditor prepare especially informative annual reports (see the discussion on reports in chapter 5, "Addressing Important Issues in Citizen Oversight").

For guidance in developing a program brochure, examine the brochures prepared by the oversight bodies in Minneapolis, Rochester, San Francisco (in English and Spanish), San Jose, and Tucson. Several jurisdictions have prepared reports recommending modifications to their existing oversight procedures. These reports provide valuable discussions of alternative approaches to citizen oversight. See, for example, the following (available from the oversight bodies):

ALBUQUERQUE, PORTLAND, AND TUCSON DID THEIR OWN RESEARCH (CONTINUED)

Tucson

In 1996, the mayor and city council of Tucson asked staff to provide information on alternative models for external police review procedures, including potential changes to the city's existing Citizen Police Advisory Review Board. Staff consulted with involved parties throughout the city and obtained additional information from a number of well-known practitioners and experts from across the Nation. The resulting 33-page report identified the limitations of the current commission, reviewed alternative models, and provided recommendations for improving the city's current procedure and cost estimates for each proposed improvement.

A mayor's and city council subcommittee, headed by an assistant city manager, conducted a nationwide review of options and discussed them before adopting its own ordinance. The assistant city manager telephoned cities and visited the independent police auditor in San Jose. The assistant had a budget and survey people assigned to her to conduct the survey. The subcommittee considered to whom the board and auditor should report, whether the board would have investigation powers, and who would supervise the auditor. There was a historical precedent for having a board because one had existed for several years. So it was natural to continue the existing board. The auditor was assigned to the city manager because the city charter puts the city manager in charge of all administration. An auditor was selected because the subcommittee felt a board was not the best avenue for citizens to bring complaints, monitor investigations, and do alternative intake because of turnover among volunteers and lack of time.

The city council debated the proposed options, including doing nothing. Three members of the council made up the public safety subcommittee, which conducted the investigation. The council debated whether an auditor was needed or only a strengthening of the existing board. There was a great deal of intense debate before the auditor model was agreed on.

- "Minneapolis Civilian Police Review Authority Redesign Report," November 1997, 29 pages, and the board's response to the report, "Response to MCPRA Redesign Report," April 1998.

- "External Police Review: A Discussion of Existing City of Tucson Procedures and Alternative Models, Report to the Mayor and Council," October 7, 1996, 36 pages plus appendixes.

The Portland Copwatch organization developed a 10-page "Proposal for an Effective Civilian Review Board" that is available from Portland Copwatch/People Overseeing Police Study Group, P.O. Box 244296, Portland, OR 97242; 503–288–3462. (See also "Albuquerque, Portland, and Tucson Did Their Own Research.")

The Minneapolis and Rochester oversight bodies can provide additional materials on conducting mediation.

Selected Publications and Reports

Bailey, Robert G. "The Re-Emergence of Civilian Review of Police: Seizing the Opportunity and Understanding the Trade-Offs." Unpublished paper submitted to the eighth annual IACOLE conference, September 1992, 8 pages. Discusses tradeoffs in different ways citizen oversight can be structured. Order from IACOLE (see address above).

Human Rights Watch. *Shielded from Justice: Police Brutality and Accountability in the United States,* New York: Human Rights Watch, 1998, 440 pages. Discusses factors that contribute to human rights violations; recommends changes in police administration to reduce police misconduct; discusses civil remedies, prosecution, and other approaches to accountability; and provides case studies of misconduct and efforts at accountability in 14 cities. Address: 350 Fifth Avenue, 34th Floor, New York, NY 10118–3299; phone: 212–290–4700; fax: 212–736–1300; e-mail: hrwnyc@hrw.org.

Luna, Eileen, and Samuel Walker. *A Report on the Oversight Mechanisms of the Albuquerque Police Department.* Prepared for the Albuquerque City Council, February 28, 1997, 159 pages plus appendixes. See the description of the report in chapter 7, "Monitoring, Evaluation, and Funding." Available from the Albuquerque City Council, P.O. Box 1293, Albuquerque, NM 87103; phone: 505–768–3100.

New York Civil Liberties Foundation. *Civilian Review of Policing: A Case Study Report,* New York: New York Civil Liberties Union, 1993, 155 pages. Describes the operations of seven oversight systems and presents policy and practice recommendations. Address: New York Civil Liberties Union, 132 West 43rd Street, New York, NY 10036; phone: 212–382–0557.

Perez, Douglas W. *Common Sense About Police Review,* Philadelphia: Temple University Press, 1994, 322 pages. Discusses the nature of police misconduct, the limits of reform, and methods of evaluating different approaches to accountability; describes three types of police review systems: internal, civilian, and monitoring; suggests methods of improving police accountability; and proposes an ideal review system. Address: Temple University Press, 1601 North Broad Street, Philadelphia PA 19122.

Sviridoff, Michele, and Jerome E. McElroy. *The Processing of Complaints Against Police in New York City: The Perceptions and Attitudes of Line Officers.* New York: Vera Institute of Justice, 1989, 56 pages. Examines police officer attitudes toward New York City's Civilian Complaint Review Board based on focus groups with 22 officers. Address: Vera Institute of Justice, 377 Broadway, New York, NY 10013; phone: 212–334–1300.

Walker, Samuel. *Citizen Review Resource Manual,* Washington, D.C.: Police Executive Research Forum, 1995, 424 pages including appendixes. Defines citizen review; discusses selected characteristics of existing procedures in the country, including sources of legal authority, organizational structure, roles and missions, board composition, disciplinary authority, and information dissemination; and includes as appendixes a variety of ordinances and statutes, rules and procedures, annual reports, and other documents. Order from the Police Executive Research Forum, 1120 Connecticut Avenue, Suite 930, Washington, D.C. 20036; phone: 888–202–4563.

Walker, Samuel. *Police Accountability: The Role of Citizen Oversight,* Belmont, California: Wadsworth, 2001. Addresses many of the same issues covered in the present report. See address and phone number for Walker in exhibit 8–1.

Walker, Samuel, and Vic W. Bumphus. "The Effectiveness of Civilian Review: Observations on Recent Trends and New Issues Regarding the Civilian Review of Police." *American Journal of Police* 11 (4) (1992): 1–26. Based on a survey of 50 oversight procedures. Describes their general features, explains the pattern of growth in citizen oversight, and discusses barriers to evaluating the effectiveness of citizen review operations.

Walker, Samuel, and Betsy Wright. *Citizen Review of the Police, 1994: A National Survey,* Washington, D.C.: Police Executive Research Forum, January 1995, 19 pages. Based on an ongoing survey of oversight bodies. Discusses the types of procedures in existence, dividing them into four models; tracks the growth of oversight procedures since 1970; identifies which types of models are most prevalent in cities of various sizes; and provides a matrix listing jurisdictions by oversight model, name of the oversight body, year established, and jurisdiction size. See ordering information above.

Individuals With Experience in Citizen Oversight of Police

The individuals identified in exhibit 8–1 are available to provide technical assistance related to citizen oversight of the police by telephone. In addition, chapter 2, "Case Studies of Nine Oversight Procedures," provides the names of program coordinators and law enforcement administrators who are available to provide telephone consultation.

EXHIBIT 8–1. INDIVIDUALS WITH EXPERIENCE IN CITIZEN OVERSIGHT OF POLICE

Name	Title or Position	Contact Information	Areas of Experience
K. Felicia Davis, J.D.	Legal consultant and director at large, NACOLE Administrator, Syracuse Citizen Review Board	Citizen Review Board 234 Delray Avenue Syracuse, NY 13224 phone: 315–448–8750/8058 fax: 315–448–8768	• Legal consultant, NACOLE • Administrator, Syracuse Citizen Review Board • Startup problems • Working with police union
Mark Gissiner	President, IACOLE (1995–99) Senior human resources analyst, Cincinnati	2665 Wayward Winds Drive Cincinnati, OH 45230 phone: 513–624–9037 e-mail: IACOLE1@fuse.net fax: 513–352–5223	• Former president, IACOLE • Staffs commission that hears appeals of disciplinary actions in Cincinnati • Designs and provides technical assistance in creating oversight systems nationally and internationally
Douglas W. Perez, Ph.D.	Assistant professor, Plattsburgh State University	Department of Sociology Plattsburgh State University 45 Olcott Lane Rensselaer, NY 12144 phone: Fri.–Mon.: 518–426–1280 Tues.–Thurs. 518–564–3306 fax: 518–564–3333	• Former deputy sheriff • Wrote book on oversight procedures • Internal affairs procedures
Jerry Sanders	Former chief, San Diego Police Department	United Way of San Diego County P.O. Box 23543 San Diego, CA 92193 phone: 858–492–2000 fax: 858–492–2014	• Police administration • Former SWAT commander • Former training academy director • Implemented community policing departmentwide
Lt. Steve Young	Vice president, Grand Lodge, Fraternal Order of Police	Fraternal Order of Police 222 East Town Street Columbus, OH 43215 phone: 614–224–5700 fax: 614–224–5775	• Union negotiator • Police officer • Certified police instructor
Samuel Walker, Ph.D.	Kiewit Professor, University of Nebraska at Omaha	Department. of Criminal Justice University of Nebraska at Omaha 60th and Dodge Streets Omaha, NE 68182–0149 phone: 402–554–3590 fax: 402–554–2326	• Studies oversight procedures • Conducted survey of early warning systems • Author of books on citizen oversight
Note: The individuals in this exhibit have agreed to respond to telephone calls for technical assistance for citizen oversight of police. The individuals are members of the project advisory board who served as advisers in the preparation of this report.			

Glossary

Burden of proof. The standard used to conclude that an officer committed the alleged misconduct. (See "Preponderance of the evidence" and "Clear and convincing evidence.")

Clear and convincing evidence. The degree of proof that will produce in the mind of the trier(s) of fact a firm belief or conviction as to the allegations; an intermediate burden of proof, being more than mere preponderance but not to the extent of such certainty as is required by beyond a reasonable doubt.

Discipline. Punishment meted out for misconduct by officers. In increasing order of severity, punishment may include:

- Verbal or written counseling by the officer's supervisor.

- Remedial training.

- Professional counseling (e.g., for substance abuse).

- Verbal reprimand (supervisor orders inappropriate behavior to be corrected).

- Written reprimand.

- Suspension without pay (may include ineligibility for promotion, department-paid health insurance premium payments, and off-duty enforcement work).

- Probation.

- Demotion (reduction in rank, job classification, or pay grade or step).

- Termination.

Exonerated. See "Findings."

Findings. The internal affairs or citizen review oversight body's determination of the legitimacy of a citizen's complaint. Options include:

- Unfounded: The alleged incident did not occur, or the subject officer was not at the scene.

- Exonerated: The incident did occur, but the officer's actions were lawful and proper.

- Not sustained: There is insufficient evidence to prove or disprove the allegations.

- Sustained: There is sufficient evidence to conclude that the officer engaged in misconduct.

- Policy failure: The officer acted incorrectly but, because the department had no policy, an ambiguous policy, or contradictory policies prescribing the correct behavior for the situation at issue, no blame is attached to what the officer did.

***Garrity* warning.** Under *Garrity* v. *New Jersey,* 385 U.S. 493 (1967), as part of an internal, noncriminal investigation, police officers can be ordered to give a statement to their employer regarding actions they took while working for the police department. A *Garrity* warning informs the officers of two conditions: (1) failure to answer questions related to the scope of their duties may form the basis for disciplining officers, including dismissal, and (2) any statements given under this warning cannot be used in any subsequent criminal proceeding against the officers unless the officers are alleged to have committed perjury in their statements.

Internal affairs. Section, unit, or bureau within a law enforcement agency responsible for investigating officer misconduct.

Not sustained. See "Findings."

Policy failure. See "Findings."

Professional standards. In some law enforcement agencies, a new name for the internal affairs unit; in other agencies, a unit that houses several activities (e.g., inservice training) for which investigating officer misconduct is only one responsibility.

Subject officer. The officer against whom a complaint has been filed.

Glossary

Subpoena power. The authority to compel witnesses to appear and give testimony or produce relevant documents.

Preponderance of the evidence. Enough evidence to decide that one party in the case has the stronger evidence on his or her (or its) side, however slight the advantage may be, even when the trier has a reasonable doubt that the party is in the right or the party's rightness is not even clear and convincing.

Suspension. Time off without pay. Suspended officers may not be allowed to work off-duty details and may have to pay their own health insurance premiums for the time they are suspended.

Sustained. See "Findings."

Unfounded. See "Findings."

Appendix A

Rochester Internal Affairs Request for Board Member Evaluations of Investigators' Investigations

---

City of Rochester - Police Department

INTER-DEPARTMENTAL CORRESPONDENCE

TO: C.R.B. Members

FROM: Lieutenant James Sheppard, Professional Standards Section

DATE: June 3, 1998

SUBJECT: P.S.S. Investigation Evaluation

C.R.B. Member:

In order to ensure that the quality of P.S.S. investigations remain at a high standard, we would like your input. Regarding the investigation you have just reviewed, please answer the following questions as explicitly as possible:

Date................................

1: Did adequate documentation exist identifying officers and witnesses at the scene of the incident? If not, what type of documentation would you suggest?

2: Were any documents contained in the investigative package inappropriately added in your opinion?

3: Was the investigation slanted in favor of the officers involved by any of the investigating supervisors in this case? If so, by whom and how slanted?

4: Did you feel that the P.S.S. finding was supported by the information contained in the investigation? If not, what should the finding have been and why?

5: Was the investigative package presented in an organized manner? If not, what was the problem with the package?

6: Was the investigation deficient in any other way? If so, how?

Appendix B

Portland Auditor's Guidelines for Reviewing Internal Affairs Investigations

MONITORING WORKSHEET

Instructions: On the back of this form prepare a summary of the case or attach a copy of the police disposition letter. Do not identify persons involved in the complaint by name -- use complainant (CO), police officer (PO), witnesses' relationship (3 backup PO's, CO's mother, CO's friend, etc.)

IAD Case Number: _____   Monitored by _____

Date Complaint Filed   _____

Date Assigned to Commander for disposition   _____

Date Returned with findings
(Please note any requests for additional time)   _____

Who issued finding   _____

Date Complaint Closed   _____

Investigator: _____

How is complaint classified? 1) Property 2) Use of Force 3) Conduct 4) Disparate Treatment 5) Communication 6) Performance 7) Procedure

What is (are) the finding(s)? A) Unfounded B) Exonerated C) Insufficient Evidence D) Sustained E) Unsubstantiated F) Information

What is the general theme of the complaint? (Such as rudeness, use of pepper spray, improper search, etc.)

Are all allegations properly identified? If not, specify: ____ Yes ____ No

Did the disposition letter adequately explain the finding?   ____ Yes ____ No
If not, explain:

Are all interviews taped? If not, what's missing?   ____ Yes ____ No

Are all tapes properly identified? ____ Yes ____ No

Are all tapes audible? ____ Yes ____ No

Who was interviewed? (no names)

Did investigator make good, balanced witness selections? Should additional witnesses have been interviewed? Explain. Were all reasonable efforts expended to identify and locate potential witnesses?

Evidence obtained (circle):
- Photographs
- Dispatch/BOEC records
- PPDS/DMV/LEDS
- Medical records (including detox, detention center)
- Police reports / duty notebook entries / citations
- Site examination
- Court records
- Other _____

Would other evidence be helpful? Were all reasonable efforts expended to obtain evidence?

Are you satisfied that the investigator pursued all issues and asked reasonable, neutral questions? <u>Support your answer with specific examples.</u>

Did written summaries accurately reflect information provided in interviews?

Would further documentation of investigative steps been helpful?

Did decision-maker have to request additional information? Were findings controverted? If so, please explain:

Does the evidence in the file support the conclusion? Did the findings address each allegation? If not, explain.

Please note the following information:

1. If case was dropped or suspended, note reason why.
2. Note and explain, if possible, any gaps of time during course of IID investigation.
3. Any Police Bureau finding in the case that you question.
4. Anything about the IAD investigation that you question.
5. Any other issues you wish to discuss.

Appendix C

San Francisco Complaint Intake Form

Police Commission for the City and County of San Francisco

OFFICE OF CITIZEN COMPLAINTS

~CITIZEN COMPLAINT FORM~

INSTRUCTIONS FOR COMPLETION OF THE CITIZEN COMPLAINT FORM:
Please answer questions in blocks 2, 3, 4, 5, 8, 11, 15, 17, 20, 21 & 22. Leave all other blocks blank unless you know the information requested. Please **print** all information in **English**. If you do not have a telephone number, enter a message number or the number of a neighbor, friend or relative in block 4. If witnesses are available, write their names, addresses and telephone numbers on a separate sheet of paper and attach it to your complaint. Do not write them on the complaint form. If you do not know the officers name or badge number, include a complete physical description in the narrative (22). **Print** your narrative. Explain what happened from beginning to end. Be specific as to the nature of your complaint against each officer. Include who, what, where, when and why. If you need additional space, use separate sheets of paper and attach them to the complaint. **YOUR STATEMENT MUST BE A TRUE AND ACCURATE ACCOUNT OF THE INCIDENT** to the best of your knowledge and belief, and must be signed by you in block 25. If you have questions or need help, please call the OCC at (415) 597-7711 between 8:00 a.m. and 5:00 p.m., or leave a message with our answering service after 5:00 p.m. You may also contact your local neighborhood center for help. Interpreters can be provided at no charge.

填寫公民投訴書說明：
請回答第2，3，4，5，8，11，15，17，20，21及22項問題。除非您知道我們所要求的資料，否則請將其他各項留空。所有資料，務請以正楷填寫清楚。如果您沒有電話號碼，請在第4項填上有可能聯絡您的電話號碼，或鄰居、親戚、朋友的號碼。如果有證人，請用另一張紙寫上他們的姓名、地址及電話，和投訴書夾在一起；切勿寫在投訴書上。如果你不知道涉及事件的警務人員姓名或編號，請將該員的身體特徵，以正楷詳盡寫在第22項上。請清楚說明事件的過程，及投訴的類別，包括涉及何人、何事、何處、何時及何由。如您認為投訴書不夠您填寫，可以另紙填寫資料，夾在投訴書上。您應根據您所知道及所相信的事實填寫資料，必須真實及正確；填妥請在第26項簽名。如有疑問或需要幫助，請在上午八時至下午五時，致電 (415) 597-7711「公民投訴組」，或在下午五時後，在該組的電話錄音機上留言。您亦可以與有免費翻譯員服務的「華埠建民中心」求助。電話415-391-5099。

INSTRUCCIONES PARA LLENAR EL FORMULARIO DE QUEJAS DE LOS CIUDADANOS: Por favor conteste las preguntas de las casillas 2, 3, 4, 5, 8, 11, 15, 17, 20, 21 & 22. Deje sin contestar las demas preguntas a menos que sepa la información solicitada. El formulario debe ser contestado en **Ingles**. Si usted no tiene telefono escriba en la casilla 4 el número de un servicio de mensajes, o el de un vecino, amigo o pariente. Escriba en una hoja separada los nombres, direcciones y telefonos de los testigos (si los hay), y adjunte ésta información al formulario. En caso de que no conozca el nombre o número de insignia de los oficiales, incluya una descripción fisica completa (22). Describa los hechos en forma completa, sea específico. Incluya quien, que, donde, cuando y porque. Su declaración debe ser un recuento exacto y verdadero del incidente y debe estar firmada por usted (25). Para pedir información o solicitar ayuda visite nuestras oficinas locales o llamenos al numero (415) 597-7711 de 8:00 AM - 5:00 PM. El servicio de interpretacion es gratis. Formularios tambien pueden ser obtenidos en La Raza Information Center-- (415) 863-0764.

PARAAN NG PAGSAGOT SA PORMANG ITO (CITIZEN COMPLAINT o REKLAMO NG MAMAMAYAN)
Mangyaring sagutin ang mga tanong sa blokeng 2, 3, 4, 5, 8, 11, 15, 17, 20, 21, at 22. Kung wala kayo ng impormasyon hinihingi dito, paki-iwanan blanko ang blokeng hindi masagot. **Paki-limbag ang lahat na sagot ninyo.** Kung wala kayong telepono, paki-sulat lang ang inyong "message number", o ang numero ng inyong kapit-bahay, kaibigan, o kamaganak. Kung mayroon kayong mga saksi o testigo, isulat sa ibang papel ang kanilang mga pangalan, mga tirahan, at mga telepono at ikabit ito sa reklamo ninyo. Huwag gagamitin ang pormang ito. Kung hindi ninyo alam ang pangalan ng pulis o ang numero ng kanyang tsapa, isama sa inyong salaysay ang hitsura at pagmumukha ng pulis. Ilimbag ang inyong salaysay. Liwanagin lahat ang nangyari magmula sa umpisa hanggang sa katapusan. Tiyakin o siguraduhin ang inyong sinusumbong o renireklamo. Sabihin o ilarawan kung sino, ano, saan, kailan at bakit sa pangyayari. Kung kulang ang pagsusulatan dito gumamit ng ibang papel at ikabit ito sa sumbong ninyo. Sa inyong kaalaman at paniniwala, ang inilahad ninyong nangyari ay dapat lubos na katotohanan at walang kamali-mali at kailangan ninyong pirmahan ang sumbong ito sa blokeng bilang 25. Itanong sa amin kung alinman dito ang hindi maliwanag sa inyo. Kung kailangan ninyo ng tulong, paki-tawagan kami, OCC, telepono (415) 597-7711. Maaring tawagan din ninyo ang Philippines American Consul sa telepono (415) 626-0773 sa pagitan ng alas--otso ng umaga at alas--singko ng hapon o mag-iwan ng pahatid o "message" sa aming "answering service" paglampas dng alas--singko ng hapon.

SFPD/OCC FORM 293

APPENDIX C

OFFICE OF CITIZEN COMPLAINTS - USE BLACK INK ONLY!

1 Day, Date & Time Complaint Received

Complaint Against: Personnel ☐ Policy ☐ Procedure ☐
How Received: Person ☐ Phone ☐ Letter ☐ SFPD ☐ Mail-In ☐ Other ☐ : (specify)_____

2 Primary Complainant: ○ Co-Complainant

Last Name _____ First Name _____ Middle Initial

HOME ADDRESS: _____
Street   Apartment

City _____ State _____ Zip

WORK ADDRESS: _____
Street   Apartment

City _____ State _____ Zip

3 Personal Information
Age: ____ Date of Birth: ____
Sex: ____
Ethnicity: ____
Occupation: ____

4 Telephone Numbers:
Home: (____) ____
Work: (____) ____

5 Location of Occurrence:

6 Type of Place   **7** District

8 Day, Date, & Time Of Occurrence: A.M. / P.M. (Circle one)

9 Incident Report or Citation No

10 SECONDARY COMPLAINANT? Yes ☐ No ☐ Witnesses? Yes ☐ No ☐ (If "Yes", attach separate sheet of paper.)
Taped Interview? Yes ☐ No ☐ Criminal Case Pending in Relation to this matter? Yes ☐ No ☐

11 Injuries Claimed? Yes ☐ No ☐ Injuries Visible? Yes ☐ No ☐ Drug/Alcohol Related? Yes ☐ No ☐
Photos Taken? Yes ☐ No ☐ By: Photo Lab ☐ O.C.C. ☐ Other: ____
Type of Injury: Medical Release Signed? Yes ☐ No ☐

12 Activity	**13** Type	**14** DISP.	**15** Uniform Yes No	**16** Rank	**17** Member's Name & Star Number	**18** Unit	**19** Svc	**20** Sex	**21** Eth

(22) NARRATIVE OF INCIDENT:_____

_____
_____
_____
_____
_____
_____
_____
_____
_____
_____
_____
_____
_____
_____
_____
_____
_____
_____
_____
_____
_____
_____
_____
_____
_____
_____

(State law passed in 1995 mandates that the following statement be provided to, read and signed by persons filing complaints. The OCC encourages the filing of a complaint by anyone who believes he or she is a victim or a witness of improper police conduct or policies.)

ACKNOWLEDGEMENT OF COMPLAINANT (148.6 P.C.)

YOU HAVE THE RIGHT TO MAKE A COMPLAINT AGAINST A POLICE OFFICER FOR ANY IMPROPER POLICE CONDUCT. CALIFORNIA LAW REQUIRES THIS AGENCY TO HAVE A PROCEDURE TO INVESTIGATE CITIZENS' COMPLAINTS. YOU HAVE A RIGHT TO A WRITTEN DESCRIPTION OF THIS PROCEDURE. THIS AGENCY MAY FIND AFTER INVESTIGATION THAT THERE IS NOT ENOUGH EVIDENCE TO WARRANT ACTION ON YOUR COMPLAINT; EVEN IF THAT IS THE CASE, YOU HAVE THE RIGHT TO MAKE THE COMPLAINT AND HAVE IT INVESTIGATED IF YOU BELIEVE AN OFFICER BEHAVED IMPROPERLY. CITIZEN COMPLAINTS AND ANY REPORTS OR FINDINGS RELATING TO COMPLAINTS MUST BE RETAINED FOR AT LEAST FIVE YEARS. IT IS AGAINST THE LAW TO MAKE A COMPLAINT THAT YOU KNOW TO BE FALSE. IF YOU MAKE A COMPLAINT AGAINST AN OFFICER KNOWING THAT IT IS FALSE, YOU CAN BE PROSECUTED ON A MISDEMEANOR CHARGE.

☐ I HAVE READ AND UNDERSTOOD THE ABOVE STATEMENT.   ☐ THE ACKNOWLEDGMENT HAS BEEN READ TO THE COMPLAINANT.

Complainant Signature/Date:	Taken By (Name/#/Unit)/Date:
Assigned Investigator/Date:	Closure Approval/Date:

APPENDIX C

After you have completed this form, return it to the Office of Citizen Complaints by folding it along the lines below so that the address shows on the outside. Drop in any mailbox. NO POSTAGE NECESSARY IF MAILED IN THE UNITED STATES.

在您填妥本投訴書後，請沿摺線摺妥（地址在外），投入郵箱，寄回「公民投訴組」。在美國境內寄出，不需郵費。

Despues de completar la forma, doblela sobre las lineas marcadas y depositela en el buzon. No necesita estampilla (sello postal).

Matapos buuin ang pormang ito, tiklupin sa mga linyang nakatatak sa baba upang makita sa labas ang aming "address". Ihulog sa anumang buson o "mailbox". Hindi kailangan ng selyo kung ipadadala lang sa loob ng America.

OFFICES LOCATED AT:
480 Second Street, Suite 100
San Francisco, CA 94107

NO POSTAGE
NECESSARY
IF MAILED
IN THE
UNITED STATES

BUSINESS REPLY MAIL
FIRST CLASS MAIL PERMIT NO. 22978 SAN FRANCISCO, CA.

POSTAGE WILL BE PAID BY ADDRESSEE

City and County of San Francisco
OFFICE OF CITIZEN COMPLAINTS
875 Stevenson Street, Room 125
San Francisco, CA 94103-0917

Appendix D

San Francisco Policy for Citizen Monitoring of Police During Demonstrations

O.C.C. Policy : Monitoring of Demonstrations

Effective Date: 1/1/98 and continuing
Issued By: Mary C. Dunlap, Director
and
Robert S. Janisse, Chief Investigator

I. PURPOSE

The mission of the Office of Citizen Complaints (below, "OCC") is "...to achieve accountability of every member of the San Francisco Police Department, in each and every rank, position and location, to all of the people in or of this City and County." (Mission Statement of the OCC, 7/29/96). Periodically, the OCC is called upon to monitor the conduct of members of the San Francisco Police Department and to observe interactions of SFPD members with civilians during mass interactions, such as demonstrations. OCC has performed this public service frequently during its existence, and to good effect, by following policies similar to those recited herein. OCC monitors record acts of possible misconduct and of questionable conduct by SFPD sworn members on duty. OCC monitors' observations can serve to document, identify, interpret and evaluate the potential merits of OCC complaints. Also, in the experience and history of the OCC, OCC monitors' presences and monitoring duties have served in the past to deter misconduct in some instances.

II. POLICY

Accordingly, it is the policy of the OCC to monitor demonstrations when it is determined to be consistent with OCC's mission, and feasible and advisable to do so, in the joint determination of the Director and Chief Investigator of the OCC, as informed by other OCC staff, members of SFPD and the public. The Director and Chief Investigator (or their designee(s), in their absence(s)), shall decide whether, when, where and how OCC monitoring shall occur.

III. NOTIFICATION

The primary source of demonstration notifications is the San Francisco Police Department (below, "SFPD"). The OCC may also receive information about pending demonstrations from community groups, bulletins, Internet communications, or from individual persons. The OCC will actively seek and will carefully evaluate information regarding demonstrations, from whatever source, and will make determinations about monitoring based on the best possible information available.

When notified about a demonstration, the Director and Chief Investigator will analyze the desirability and feasibility of monitoring the event, guided by the Mission Statement of the OCC and the OCC's past experience and cumulative expertise as to monitoring demonstrations. The analysis will consider all available information concerning (but not limited to) the following variables: probable size, constituencies, past relevant behavior, and other pertinent experience as to involved groups and individuals; and, any other information that logically affects the issue of OCC monitoring.

When it is determined that OCC monitoring shall occur as to a specific event or group of events, a Demonstration Notification will be posted and/or circulated at the OCC. Trained OCC staff may then volunteer for assignments as needed, to be compensated consistent with all relevant legal, Civil Service and MOU requirements. In the absence of volunteers, the Director and Chief Investigator shall make equitable monitoring assignments, if they choose, based on the needs of the OCC and the public, with due consideration to all individual staff commitments.

SFPD event operations orders and other planning information shall be provided to the Director and Chief Investigator or their authorized designees upon request. The need for OCC to know SFPD staffing and operations plans is critical to the effective deployment of OCC monitors. Any staffing or operations plans or related information received from SFPD shall be considered confidential by all informed OCC staff members and is not to be disclosed, absent valid legal compulsion (e.g. a subpena or other legally enforced requirement of testimony or production of documents), to any person outside the OCC staff, whether before, during or after an event.

III. DUTIES OF OCC MONITORS

OCC monitors shall at all times be supervised by a person or persons designated by the Director and Chief Investigator to supervise OCC monitors. Monitors shall at all times obey the lawful directions of their monitor supervisors. In following the directions of their monitor supervisors, monitors shall also follow all of the guidelines recited below to the extent practicable; in the event of conflict between a supervisor's direction and one or more of these guidelines, the monitor shall do his/her best to obey the supervisor's direction:

1) <u>The safety of monitors shall at all times dictate their location.</u> Monitors shall be placed, or shall place themselves, in positions that will afford them the best safe view of the overall demonstration.
2) Monitors shall make every reasonable effort to remain visible to SFPD members, demonstrators and members of the public.
3) Monitors shall not routinely or without reason place themselves in the midst of demonstrators, nor on the ground between police and demonstrator lines or groups.
4) Monitors should seek observation positions above and behind SFPD and/or demonstrators, consistent with the above guidelines, where those positions afford an overall view of the demonstration.
5) If and when it becomes necessary to approach and contact a police officer, a demonstrator, or any other person, the monitor shall only do so when it appears safe to do so and when it appears unlikely to interfere with any ongoing police action. All such contacts shall be brief and businesslike, so that monitors may return to their observation position as quickly as possible.
6) If an incident of possible SFPD misconduct (including neglect of duty) is observed by monitors, they shall record as much information about the incident as possible, including the date, time, location, officer identification, complainant information, and any other information that can be gathered from the observation position and that is normally sought in investigating such incidents. The use of

written notes and audio recorders is encouraged whenever feasible for these purposes.

7) If an OCC monitor is contacted by a civilian who wishes to make an OCC complaint while the event is ongoing, the monitor shall provide the person with an OCC Complaint Information Card, and, when appropriate, with a business card. After the OCC monitor has gathered as much information as possible about the particular complaint of the person, the person should be requested to contact the OCC during business hours (and generally at the earliest opportunity on the next business day) for follow-up.

8) Whenever an incident(s) of continuing serious misconduct by one or more SFPD members, or of misconduct by one or more SFPD members which results in serious injury to one or more demonstrators, is/are observed, the OCC monitor shall report the incident(s) immediately to his/her assigned monitor supervisor. The monitor supervisor shall then seek to locate and contact the ranking on-scene command officer in order to provide a report of the incident(s) to the on-scene SFPD command officer, so that the on-scene command officer can take appropriate action, including but not limited to reassignment of the officer(s) involved in the incident(s).

9) <u>OCC monitors shall not be placed and shall not place themselves in risky locations for the purpose of improving their ability to observe.</u> They shall avoid proximity to any demonstrator action or police action that poses a hazard to their safety. If objects are being thrown or items set on fire, monitors shall immediately vacate the hazardous area. In the event of fires, monitors shall be mindful of wind direction to avoid smoke inhalation.

10) If an OCC monitor observes an illegal or unlawful action on the part of one or more demonstrators, which action is likely to cause injury to anyone, the OCC monitor shall notify her/his OCC monitor supervisor immediately, or, in the instance that an OCC monitor supervisor cannot be contacted and informed, they shall notify the nearest SFPD command officer. OCC monitors witnessing and reporting any such civilian activity shall not submit to field interviews unless their OCC monitor supervisor tells them to do so, or, in the absence of an ability to contact their OCC monitor supervisor, the seriousness of the situation requires a field interview. Instead, OCC monitors shall provide their name and a contact telephone number (preferably by means of an OCC business card) in order to facilitate a later interview.

11) <u>At no time shall any OCC monitor or monitor supervisor take any action that would interfere with any law</u>

enforcement officer or group of law enforcement officers engaged in the performance of their duties at a demonstration.

12) Monitors should not break the integrity of a police line, barricade or arrest encirclement. If a situation in need of OCC monitoring develops, monitors shall contact their monitor supervisor, who shall contact the ranking command officer at the scene for assistance in gaining access to closed or cordoned-off areas. If a monitor is in any danger of physical assault at a demonstration, the monitor shall make every reasonable effort to gain the attention and assistance of SFPD members.

13) OCC monitors and monitor supervisors shall avoid conflicts and confrontations with members of SFPD at demonstrations. OCC monitors and monitor supervisors shall follow orders and directions from SFPD members to the fullest extent possible consistent with OCC monitors' personal safety. Should a conflict or confrontation arise with any officer that affects an OCC monitor's ability to monitor a demonstration, that conflict or confrontation should be carefully documented by the OCC monitor and should be reported immediately to the OCC monitor supervisor. After the termination of the demonstration, a memorandum should be prepared and submitted to the monitor supervisor, or, if prepared by the monitor supervisor, to the Director and Chief Investigator. It shall be the Director's responsibility to take appropriate action to resolve any reported conflicts or confrontations between OCC monitors and SFPD members at demonstrations.

14) OCC monitors and monitor supervisors shall wear at least one item of OCC-identified clothing at demonstrations. OCC monitors must have their badge and employee ID cards in their possession at demonstrations. Monitors should be aware that their recognition by others at demonstrations is important, and act accordingly.

15) No less than two OCC monitors (including a monitor supervisor) will be assigned to any one demonstration, except that a single staff member may be dispatched to the scene of an event to help ascertain the need for OCC monitoring. Should monitoring be needed, the staff member shall notify the OCC Director or Chief Investigator immediately.

16) OCC monitors are non-partisan observers, and do not participate in demonstrations while assigned as monitors to demonstrations. OCC monitors shall dress and act accordingly.

Appendix D

17) If an OCC employee is injured while on-duty in any capacity, the employee or another OCC employee shall notify the OCC monitor supervisor immediately. The injured employee shall be immediately transported to the facility of her/his choice, or to the nearer facility if he/she cannot choose, as follows:

The San Francisco General Hospital Occupational Health Service Clinic, 1001 Potrero Avenue, Building 9, 2nd Floor (free designated parking is available to injured employees seeking treatment at this clinic by entering the SFGH parking lot at 22nd Street and following the signs to the OHS parking lot); clinic hours are Monday through Friday, 7:30 a.m. to 5:00 p.m.

The Mount Zion Occupational Medicine Clinic is located at 1515 Scott Street, between Geary and Post (paid parking available at this site will be reimbursed by workers' compensation); clinic hours are Monday through Friday, 8:00 a.m. to 5:00 p.m. (phone: 885-7770)

All employees who require treatment after hours or on the week-end should report to the **UCSF/Mt. Zion Emergency Departmnt** which is located on Sutter Street, between Divisadero and Scott Streets. (phone: 885-7520)

If an OCC employee is too seriously injured to go to one of these locations, an ambulance should be summoned.

18) Upon termination of the demonstration or at the end of monitoring duties as defined by the OCC monitor supervisor, if during OCC business hours, the OCC monitor shall return to the OCC office. If not during OCC business hours, the OCC monitor shall only return to the OCC office at the direction of the on-scene OCC monitor supervisor. As soon as practicable, and no later than the end of the first business day after the demonstration, the OCC monitor shall provide a report <u>if she/he witnessed acts which appear to him/her to have constituted possible SFPD misconduct or neglect of duty</u>. The report shall contain the following: one memorandum per incident separately identifying and detailing the incident of possible police misconduct that the OCC monitor observed, containing all observations made by the OCC monitor. (See part 6, above). Any documentation (e.g. audiotapes) should be attached or coherently referenced in the memorandum. Two copies of all memos shall be provided by the OCC monitor to her/his OCC

monitor supervisor by no later than the end of the first business day after the demonstration that he/she monitored. 19) The OCC on-scene monitor supervisor shall prepare two copies of an "OCC Monitoring Report" for each demonstration monitored, attaching a copy of each memo from each OCC monitor (including herself/himself) witnessing an incident of possible SFPD misconduct. The two copies of the "OCC Monitoring Report" shall be provided by the on-scene OCC monitor supervisor to the Director and Chief Investigator by the end of the second business day after the demonstration. 20) OCC personnel shall conduct themselves with due regard for OCC policies of courtesy, professionalism and consideration of others when monitoring demonstrations.

V. COMMUNICATIONS

OCC monitor supervisors, or their designees, should carry PIC radios at demonstrations, to monitor the radio traffic for demonstration and police deployment information. When needed, and where feasible, additional PIC radios shall be borrowed from SFPD in order to facilitate OCC monitoring of demonstrations. Cel/portable telephones, and not PIC radios, should be used for communication between or among OCC monitoring teams.

VI. SUMMARY OF OCC POLICY ON MONITORING DEMONSTRATIONS

OCC shall monitor demonstrations when it is decided by OCC's Director and Chief Investigator that it is consistent with OCC's mission, and feasible and advisable for OCC to do so. All OCC monitoring of demonstrations shall be conducted by designated OCC staff members in accordance with the terms of this policy and of the training of OCC staff members as to monitoring of demonstrations. In case of any difficulty, conflict or question, OCC monitors shall consult with their OCC monitor supervisors, and OCC monitor supervisors shall consult with OCC managers including, where appropriate, the Chief Investigator and the Director of the OCC.

Appendix E

Cost Estimates for Different Modifications to Tucson's Existing Oversight Procedure

External Police Review:
A Discussion of Existing City of Tucson Procedures and Alternative Models

Many factors will impact the ultimate cost of providing these review functions, including:
- types of complaints to be received, investigated or reviewed
- extent of mediation and conciliation to be attempted
- review body's method of obtaining legal services
- method of providing professional investigations (staff vs. contract).

The following are preliminary projections for added costs associated with providing the alternative function. Costs associated with obtaining any required charter or police union contract changes are not included.

Expand current review functions to include an intake function
How an intake function is established will affect its cost. Intake can be as simple as taking information over the phone. Considering an estimated 800 annual citizen complaints, staff projects that providing an expanded intake function would require one clerk, one-half of a supervisor and recurring operating costs. The following estimates assume office space is available and certain large office machines (copier, fax) can be shared.

        Estimated total startup cost $19,000   Estimated total recurring cost $55,000
                      Estimated total first year cost $74,000

A more complex form of intake involves conducting interviews with complainants to begin the analysis required for complaint classification and determining mediation potential. Providing this function would require clerical support, a management analyst, one-half of a supervisor and recurring operating costs.

        Estimated total startup cost $25,000   Estimated total recurring cost $97,000
                      Estimated total first year cost $122,000

Expand review functions to include review of completed Internal Affairs' investigations of individual officers (includes expanded intake function)
In order to efficiently review completed investigations, it would be prudent to have the initial intake function be as thorough as possible by conducting complainant intake-interviews as opposed to simply reporting the information. It is assumed that not all of the completed Internal Affairs investigations would require a review. Some cities have given their review bodies the authority to review completed investigations, but do not provide staff for this purpose; the reviews are conducted by members of the review body. If the City were to follow this model, the costs are projected to be the same as providing the intake function:

        Estimated total startup cost $25,000   Estimated total recurring cost $97,000
                      Estimated total first year cost $122,000

External Police Review: A Discussion of Existing City of Tucson Procedures and Alternative Models, Report to the Mayor and Council, Tucson, Arizona, October 7, 1996, pp. 30-31.

APPENDIX E

External Police Review:
A Discussion of Existing City of Tucson Procedures and Alternative Models

Providing staff to conduct the review work would require clerical support, two management analysts, one-half of a supervisor and recurring operating costs.
    Estimated total startup cost $31,000   Estimated total recurring cost $142,000
    Estimated total first year cost $173,000

Conduct investigations of individual officers (includes expanded intake function and review of completed Internal Affairs investigations)
Conducting investigations independently of Internal Affairs provides the highest level of authority to an external review body. Therefore, it is logical that it is also the most costly. Some cost containment may be achieved through contracting of services, or negotiation of pro-bono services. If all staffing were in-house, providing this function would require clerical support, a management analyst, two investigators, a supervisor (Chief Investigator) and recurring operating costs, including office space rental. Contracting out the investigative services would require the supervisor and intake staff positions plus the negotiated contract for the investigations.
    Estimated total startup cost $50,000   Estimated total recurring cost $231,000
    Estimated total first year cost $281,000

Subpoena power
Subpoena power has been granted to other cities' external review bodies along with any or all of the previously mentioned powers. With use of subpoena power, an external review body can generally anticipate a large number of court challenges. How these challenges are handled will determine the financial impact. No estimate is provided for the legal costs that may result from the use of subpoena power.

Independent Police Auditor Model
The functions of complaint intake and review and investigation of complaints against individual officers can be assigned to an Independent Police Auditor. Staffing under this model would require an independent auditor, a complaint intake position, clerical support, and other non-salary operating costs.

    Estimated total startup cost $50,000   Estimated total recurring cost $157,000
    Estimated total first year cost $207,000

Index

A

added allegations 58, 113, 115

Albuquerque 12, 88, 118, 128, 130, 138

annual reports 103–104, 138

appeals 24, 34, 39, 54, 65

associations (see unions)

audiences, for the publication 2–3

auditors 6, **41–46**, **62–65**, 66–68, 104–105, 147–149

B

benefits (potential) of citizen oversight **6–12**
    to communities 12
    to complainants 7–8
    to elected and appointed officials 10–12
    to subject officers and deputies 10, 115, 120
    to police and sheriff's departments 8–10

Berkeley 8, 9, 11, 12, 15, 17, 20, **21–26**, 70–71, 80, 82, 84, 84–85, 85, 87, 90, 94–96, 97, 98, 99, 100, 102, 103, 105, 112, 116, 136

board members 21–25, 46–51, 51–55, 65–68, **84–88**, 101, 104–105, 112–114, 125

Boise 6, 118

budgets (see funding)

C

case studies (see also cases, vignettes of) **17–68**

cases, vignettes of 23, 24, 29, 35, 38, 78, 79

Citizen Review of Police
    features 4–5, 18, 19
    organization of publication 2, 17
    purposes 3–4
    sources of information used in preparation of report 4–5
    terminology used in report 5, 143–144

commissioners (see board members)

community policing 9–10

complainants—eligibility 98–99

consumer satisfaction surveys **125–127**

costs (see funding)

crowd control 71, 82

conflicts **107–121**
    among local government officials 104–105
    between oversight and police and sheriff's departments 3–4, 101, 105, **109–117**
    between oversight and activists 3–4, 109

D

delays in case processing 98, **101–102**, 113–114

directors 90

discipline of officers 13, 24, 28, 34, 48, 54, 55, 59, 60, 64, 80, **111**, **143**

E

early warning systems 35, 60, **80–82**

effectiveness of citizen oversight (see also evaluation) 12–16, 123

evaluation 127–128, 130

F

federations (see unions)

filing locations 96–97

Flint 10, **26–30**, 71, 100, 104, 111, 124, 134

funding
    average cost per complaint 131–132, 136
    budgets 18, 25, 30, 36, 40, 45, 50, 54–55, 60–61, 65, **128–129**, **131–135**
    determining funding needs 84–85, 134, 136, 163–164

Citizen review programs are listed by jurisdiction, not by official title (e.g., "Rochester," not "Civilian Review Board.")

Index

G

Garrity decision 100, 143

goals of citizen oversight 109, 127–128

H

hearings, vignettes of 23, 24, 35, 38

help with setting up citizen oversight **137–141**
    experts 26, 30, 36–37, 40, 46, 51, 55, 61, 68, **140–141**
    organizations **137**
    program materials **137–139**
    publications and reports **139–140**

I

internal affairs 3, 10, 13, 43, 44, 58–59, 71, 97, 109–111, 120, 129, 143, 145–146, 147–149

International Association for Civilian Oversight of Law Enforcement 87–88, 88, 98, 119, 137, 141

investigators, citizen **88–90**, 112–114, 124–125

intake 124

K

Kansas City 89

L

legal basis for citizen oversight 98

limitations to citizen oversight 12–15

Los Angeles County 11

M

media involvement 28, 30, 45–46, 96

mediation 20, 35, 49, 60, 64, **72**, **74–80**, 105

Minneapolis 9, 11, 13, 17, 20, **30–37**, 72, 74–76, 78–79, 80, 81, 84, 85, 86, 87, 88, 89, 90, 94, 96, 98, 100, 102, 105, 112, 113, 116–117, 117, 119, 125–127, 135

models of citizen oversight 6, 17–18, 134

monitoring **124–127**

N

National Association for Civilian Oversight of Law Enforcement 137, 141

New York City 12–13, 77, 125

O

objectives of citizen oversight (see goals of citizen oversight)

Omaha 86, 120

ombudsman (see Boise, Flint)

openness of procedures 19, 20, **103–104**

Orange County 10, 20, **37–40**, 70, 71–72, 81, 85, 86, 86–87, 87, 88, 90, 96, 97, 99, 100, 101, 103, 111, 114, 115, 117

outreach **94–97**, 151–154

P

panelists (see board members)

planning checklist xiii

policy and procedure recommendations 10, 18, 20, 25, 39, 43, 45, 49, 60, 66, 69–72

politics 30, 36, **104–105**

Portland 10, 17, **41–46**, 71, 72, 73, 80, 81, 90, 98, 100, 102, 104–105, 105, 114, 115, 116, 119, 120, 138, 147–149

public forums 12, 24–25, 26, 34–35, 65–66

R

resources (see help with setting up citizen oversight)

retaliation for filing a complaint 95

ride-alongs 88

Rochester 9, 20, **46–51**, 71, 74, 77–78, 79, 80, 84, 85, 86, 87, 88, 101, 102, 103, 105, 112, 117, 119, 124, 125, 129, 145–146

Citizen review programs are listed by jurisdiction, not by official title (e.g., "Rochester," not "Civilian Review Board.")

S

St. Paul 10, 11, 17, 20, **51–55**, 72, 84, 85, 86, 87, 88, 99, 101, 111, 119

San Diego 101

San Francisco 9, 14–15, 17, 20, **55–62**, 70, 77, 80, 82, 83, 89, 90, 94, 96–97, 98, 99, 101, 102, 112, 114, 115, 116, 119, 120, 121, 124, 135, 151–154, 155, 161

San Jose 67, 104

staffing (see also board members; directors; investigators, citizen) 18, 20, 83–91

subpoena power 19, 39, **99–101**, 111, 118–119

T

terminology, used in the report 5, 143–144

tradeoffs in choosing models 6–7, 20

Tucson 17, 20, **62–68**, 70, 80, 84, 87, 96, 97, 98, 113, 114, 115, 116, 120, 127, 136, 138, 163–164

U

unions 33, 34, 54, 58, 59, 60, 64, 86, 101, **117–120**, 141

use of force 9, 16, 70, 71–72, 82, 112

Citizen review programs are listed by jurisdiction, not by official title (e.g., "Rochester," not "Civilian Review Board.")

www.ingramcontent.com/pod-product-compliance
Lightning Source LLC
Chambersburg PA
CBHW081119170526
45165CB00008B/2497